Workshop on Surface Modifications and Functionalization of Materials for Biomedical Applications

AA008586

(BIO-COAT 2010)

Journal of Physics: Conference Series Volume 252

Saragossa, Spain
24 June 2010

ISBN: 978-1-61782-127-1
ISSN: 1742-6588

Printed from e-media with permission by:

Curran Associates, Inc.
57 Morehouse Lane
Red Hook, NY 12571

Some format issues inherent in the e-media version may also appear in this print version.

Printed by Curran Associates, Inc. (2010)

For permission requests, please contact the Institute of Physics
at the address below.

Institute of Physics
Dirac House, Temple Back
Bristol BS1 6BE UK

Phone: 44 1 17 929 7481
Fax: 44 1 17 920 0979

techtracking@iop.org

Additional copies of this publication are available from:

Curran Associates, Inc.
57 Morehouse Lane
Red Hook, NY 12571 USA
Phone: 845-758-0400
Fax: 845-758-2634
Email: curran@proceedings.com
Web: www.proceedings.com

TABLE OF CONTENTS

Journal of Physics: Conference Series **252** (2010) 011001 doi:10.1088/1742-6596/252/1/011001

Preface

This special issue contains selected papers which were presented as invited and contributed communications at the workshop entitled "Surface modification and functionalization of materials for biomedical applications" (BIO-COAT 2010) which was held on June 24th 2010 in Zaragoza (Spain).

The surface of a material plays a major role in its interaction with the biological medium. Processes related to the mechanical stability of articular devices in contact, osseointegration, thrombogenicity, corrosion and leaching, or the inflammatory response of rejection of a material, are clearly conditioned by the surface properties. Therefore, the modification or functionalization of surfaces can have an important impact on these issues. New techniques for functionalization by thin film deposition or surface treatments help to improve superficial properties, while understanding the interaction of the surface-biological medium is critical for their application in new devices.

Jointly organized by the Spanish Materials Research Society, BIO-COAT 2010 provided an open forum to discuss the progress and latest developments in thin film processing and engineering of biomaterials.

Invited lectures were particularly aimed at providing overviews on scientific topics and were given by recognized world-class scientists. Two of them have contributed with a proceedings article to this selected collection (articles 012001 and 012008). The contributed communications were focused on particular cutting-edge aspects of thin film science and functionalization technologies for biomaterials, showing the major scientific push of Spanish research groups in the field. The 2010 BIO-COAT conference was organized along four main topics: (1) functionalization and texture on surfaces, (2) tribology and corrosion, (3) the surface modification of biomaterials, and (4) surface–biological environment interactions.

The papers published in this volume were accepted for publication after peer-review as are regular papers. The editor of this proceedings volume gratefully acknowledges all referees for their valuable work sometimes working to quite short deadlines.

Finally, BIO-COAT 2010 would not have been successful without the strong involvement and input of the local organizing committee in Zaragoza, and the support of the University of Zaragoza. We sincerely thank them all for their efforts.

Jose L. Endrino (Editor)
Jose A. Puértolas (Chairman)
Jose M. Albella (Chairman)

Plasma processes and applications in NanoBiotechnology

F Rossi[1] and P Colpo
European Commission, Joint Research Centre, Institute for Health and Consumer Protection (IHCP), I-21020, Ispra (VA) Italy.

E-mail: francois.rossi@jrc.ec.europa.eu

Abstract: Nanostructured surfaces presenting chemical or topographical patterns are now being increasingly developed in nanobiotechnology. Major applications are related to cell culture models and biodetection. We show that plasma technologies, in particular the combination plasma polymers deposition and etching, together with colloidal lithography, e-beam lithography and microcontact printing, are essential tools to produce nanostructured surfaces. We show that chemical and topographical patterns can be obtained on different substrates, with dimensions down to some 10 nm. The applications of these nanostructured surfaces in biology and bio-detection are reviewed and the advantages and limitation of the techniques underlined.

1. Introduction

In NanoBiotechnology, the development of bio/non-bio interfaces has been the focus of attention from the research community. It has been demonstrated that patterned surfaces with contrasted physico-chemical functionalities –e.g. adhesive/non-adhesive, hydrophilic/hydrophobic, positive/negative- are particularly interesting in tissue engineering [1,2,3,4,5], cell behavior investigations [6,7,8,9,10,11], or microfluidics [12,13]. The micro or nanostructures lead to special properties related either to the kinetics of biochemical reactions at the interface [14,15], or control of biological responses from living cells [1,5,10,16,17]. Numerous methods have been developed for the production of surfaces with controlled chemical functions using Self Assembled Monolayers (SAMs) [18,19] or electrodeposition of functional polymers [20]. The patterns can be produced by different types of lithography as well as microcontact printing [16,21,22,23]. The surface chemistries obtained are well controlled but have several limitations particularly in relation to the substrate that has to be used, for instance Gold or Silicon in the case of SAM. This is why other possibilities have been investigated, in particular in the use of functional plasma polymers that can be deposited on virtually any substrate and at low temperatures. The plasma processes allow, by the choice of the right precursors, to deposit materials a very large contrast of properties and chemical functionalities, such as –COOH [24], $-NH_2$ [25,26] and their combination [27] or anti fouling polymers based for instance on polyethylene oxide (PEO) [28,29]. An excellent review of the different functionalities produced up to now can be found in the article published by Shyong Siow *et al.* [30]. Due to their variable degree of cross-linking related to the discharge conditions, these coatings are resistant to classical polar solvents and can be used in standard lift-off processes [32,39], but can also be selectively dissolved by the appropriate solvent. Plasma polymers can be then assembled in structures with sub-micron dimensions by

[1] To whom any correspondence should be addressed

combination of colloidal lithography deposition and etching [31,32], and by direct modifications by electron beam irradiation [56]. We present here some examples of the work we performed, based on the use of plasma assisted deposition and plasma etching of functional polymers and colloidal and electron beam lithography together with microcontact printing. Application of micro-contact printing onto non-fouling PEO-like plasma polymer is also presented.

2. Experimental details

In this work, two different types of plasma polymers were used: acrylic acid plasma polymer (pAA) containing COOH moieties and plasma polymerized Polyethylene Oxide (PEO-like). The PEO-like polymer is resistant to protein adsorption in liquid solution.

2.1. Films processing

The RF plasma deposition on Si substrates of the plasma polymerized PEO-like and pAA films are described in details elsewhere [24,32,33,34]. Briefly, the systems used are capacitively coupled reactors with two symmetrical parallel-plate electrodes. For PEO-like coating, plasma polymerization is carried out by using a radio frequency generator (13.56 MHz) in pulsed mode (time on = 10 ms, time off = 100 ms, nominal power = 5 watts) of pure Diethylene Glycol Dimethyl Ether (Di-glyme), ($(CH_3OCH_2CH_2)_2O$) vapors (Sigma Aldrich, used as received). For pAA, acrylic acid vapours are mixed with Ar in pulsed RF mode (13.56 MHz; pulsed plasma: time on = 4 ms; time off =36 ms). The films obtained are uniform and their thickness can be adjusted between 20 and 200 nm. Characterisation of the pAA and PEO-like films can be found in [24,28]. Typically, the pAA films contain varying degree of carboxylic functions, depending on the power applied during the discharge. As discharge power is increased, the O/C ration as measured by XPS decreases from 50% down to a few percent, corresponding to an increase of the films cross linking and stability. For PEO-like films, similar results are obtained and the antifouling properties of the films are directly related to the importance of the C-O-C peak in the C1S XPS spectrum, which decreases as the discharge power increases [28,35]. So the composition of the films, as well as the degree of cross-linking, vary continuously as the discharge power is increased.

2.2. Surface patterning using colloidal lithography

The patterning of a surface by colloidal lithography consists in creating a 2-D crystalline structure of polystyrene (PS) beads on the pAA, those beads being used as masks during subsequent etching and deposition operation [36,37]. First, the pAA layer (thickness = 200 nm) is deposited on a Si substrate as described above. Then, a micro-drop (5 μl) of PS beads (typically of 200 to 500nm diameter) suspension is deposited on the samples surface. The sample is then spun to produce a slow evaporation of the liquid. It is thus possible to create macroscopic homogeneous areas covered by mono-layered nano-beads of few mm^2 with a surface coverage ranging from 70% to 100%. The next operation consists in a O_2 plasma etching which removes the pAA between the beads (Figure 1a). Once the polymeric nano-structures are created on the surface, the chemical contrast is formed by the plasma deposition of the PEO-like coating deposited on the surface through the nano-masks (Figure 1c). The deposition is isotropic and the regions unprotected by the masks are coated.

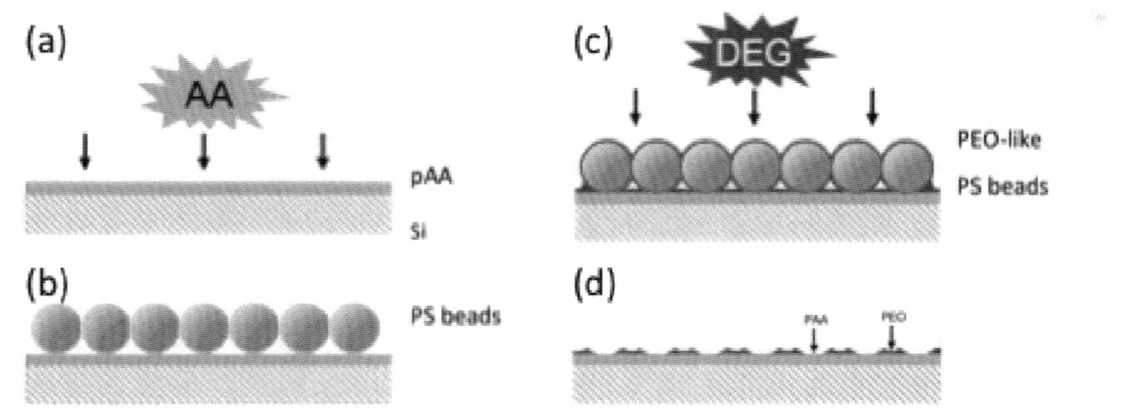

Figure 1. Process for colloidal lithography: a. Plasma polymer deposition of pAA; b. Transfer of PS beads on the substrate; c. Plasma deposition of PEO-like plasma polymer; d. Nanoparticles lift-off by ultrasonication. (AA : acrylic acid, DEG: diethylene glycol dimethylether).

The nano-sphere mask is then removed by ultrasonication in ultra-pure water. The final surface is composed of carboxylic domains of a few hundreds of nm embedded in the anti-fouling background (Figure 1d). The SEM images of Figure 2 show the surface of the sample before lift off of the beads, for different etching times. It can be seen that the patterns have an hexagonal distribution, the dimensions of which (typically a few hundreds of nm) depend on the initial size of the beads (here 500nm).

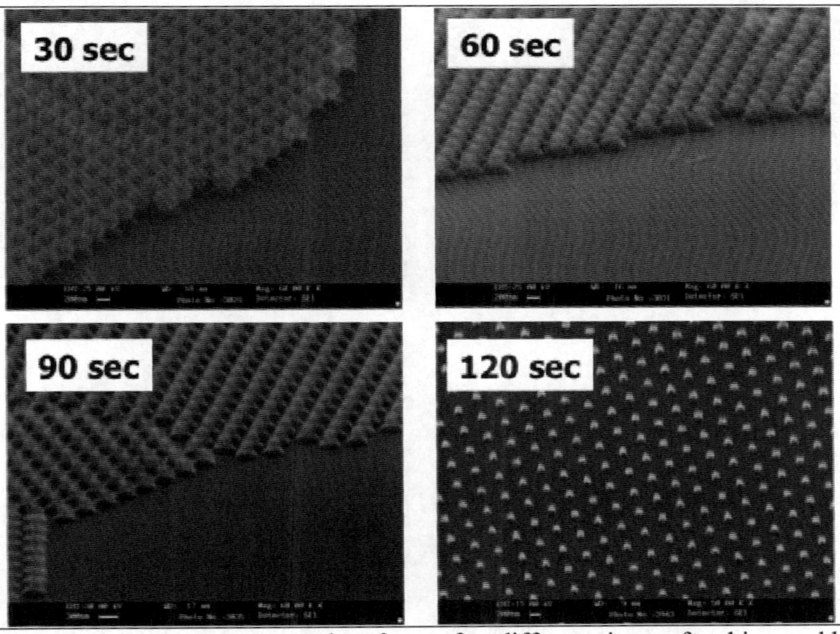

Figure 2. SEM pictures of the nanostructured surfaces after different times of etching and before lift off of the PS beads.

Moreover, the distance between active areas can be adjusted by the duration of etching. Structures of 200nm up to 1000nm and more can be obtained with this process, with coverage of about 90% over a few mm^2. Details of the experiments are given in [54,55].

2.3. Direct surface patterning using electron-beam lithography

It has been shown that plasma polymers can be patterned by direct photolithography using photoresists and PMMA [38,41]. Sub-micron nanopatterns of typically 200nm could be obtained by this approach. Lower dimensions necessitate electron-beam lithography, which can offer a high-resolution definition [39,40,41,42]. In our work, we used the possibility of cross-linking water-soluble pAA directly by the electron beam. The deposition of this layer is made with a low power (nominal power of 10 watts, operating in pulsed mode with a t_{on} of 4ms and a duty cycle of 10 %). The film thus obtained is not cross-linked and dissolves readily in water. The surface is then exposed to the electron-beam with an energy of 20 keV and incident dose of 1600 mC/cm^2, which produces a local cross linking of the polymer. After immersion in water and drying, the cross-linked domains remain on the surface. The chemistry of the film produced [56] is similar to plasma polymerized acrylic acid layer produced at optimum plasma density and used for bio specific applications [43] with a percentage of retention of COOH groups around 7-10 %.

Figure 3 shows a SEM image of pAA-structures on a PEO-like background (right) and fluorescence image of microscale structure after incubation with BSA-FTIC. Well-defined features sizes down to 200 nm in lateral resolution can be fabricated with several heights (from 80 nm to 500 nm). More experimental details can be found in [56].

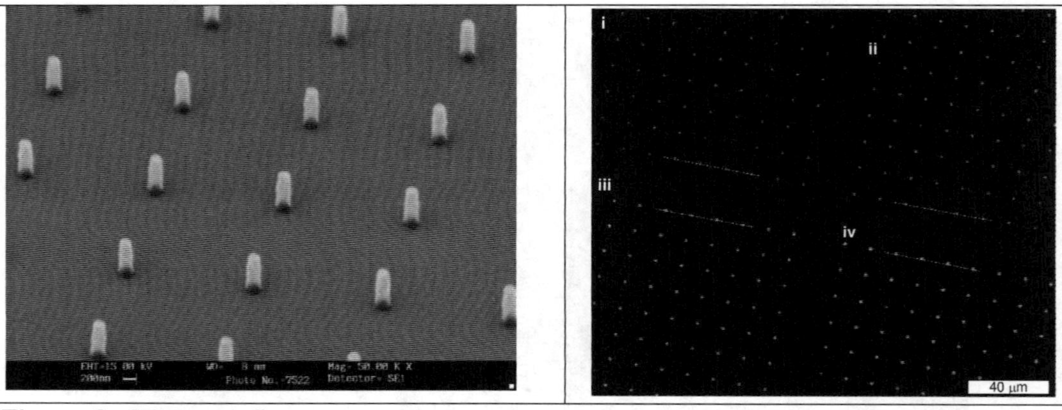

Figure 3. SEM and fluorescence picture of nanostructures obtained by e-beam patterning of an unstable pAA film and after incubation with BSA-FTIC.

2.4. Microcontact printing

On bio/non-bio interfaces, the interactions with cells or tissues are always mediated by protein adsorption on the surface [44,45,46,47]. In our work, we have optimized a platform for the patterning of human umbilical cord blood - derived neural stem cells in order to study cellular maintenance and differentiation when interacting with polypeptide micropatterns [48,49,50,51,52]. To produce these surfaces, microcontact printing has been used as a mean to deposit controlled, geometric patterns of Poly-L-lysine (PLL) and fibronectin (FN) on glass substrates previously coated with a protein resistant, cell repellent PEO-like film. The PEO-like layer has the important characteristic of being cell repellent in solution but protein adhesive under the dry conditions used during the microcontact printing operation [53]. It must be underlined that microcontact printing can be used to produce patterns down to 100nm scale.

3. Applications and results

3.1. Protein adsorption on nanopatterns

The nanostructures produced by colloidal and electron-beam lithography have been extensively studied with different biological tests and protein adsorption experiments [15,54,55,56,57,58]. In

particular, ELISA tests and Quartz Crystal Microbalance with dissipation (QCM-D) (Qsense, Sweden) has been used for quantitatively compare the mass of proteins absorbed on uniformly functionalized and nano-structured surfaces.
Both the probe (human IgG) immobilisation and the antibody/antigen (human IgG/Anti-human IgG) reaction have been monitored on-line (IgG concentrations ranging from 1 to 15 µg/ml in PBS solution). Figure 4 shows that the nano-patterned ND platform presents much higher reaction efficiency than all the other cases, evidencing the capability of the adhesives/anti-adhesives nanopatterned surface to immobilize the molecules in a reactive state, thus increasing their possibility to form complexes. This result has been further confirmed by other techniques (SPRi, QCM) [54,55,58,59,60,61] and we found that the reaction enhancement was increased as the nanostructure dimensions decrease (see Figure 5).

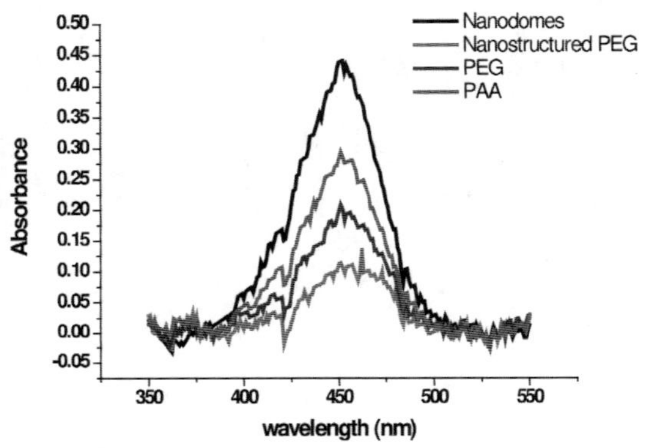

Figure 4. ELISA test performed on differently functionalized surfaces with an Ova-c-HRP concentration of 10 µg/ml

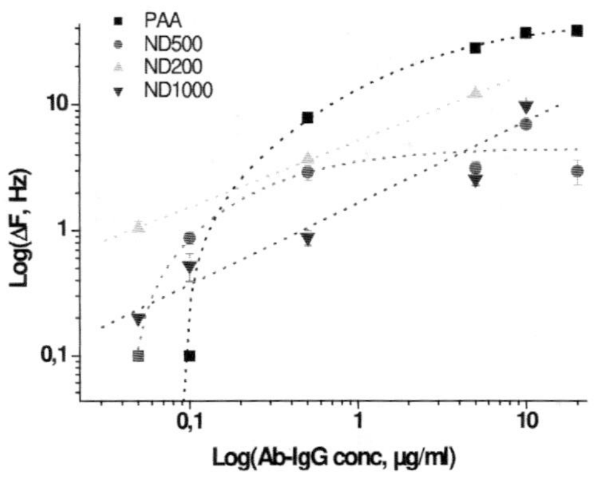

Figure 5: Calibration Curves of the immuno-reaction IgG/Ab-IgG) for the flat ppAA and the nanostructured surfaces (200, 500 and 1000nm)

A possible explanation is linked to a better orientation of the antibodies on a nanostructured surfaces and a reduction of the steric hindrance linked to the antibodies adsorption on the edges of the structures, making binding reactions more efficient on the nanostructures as compared to uniform surfaces [58].

3.2. Cell culture experiments

Microcontact printing of Fibronectin (FN) and poly-L-Lysine (PLL) was used to study the growth of neural stem cells and as a model to study different processes regarding stem cell neural lineage commitment. Human umbilical cord blood derived neural stem cells (HUCB-NSC) [62] have been incubated on the polypeptide functionalized surfaces. Previous studies on non-patterned surfaces [63] showed that the density of plating and the type of polypeptide used for attachment can change the HUCB-NSC commitment to a specific lineage (neuronal, astrocytic, olegodendroglial). In this work, the cells attached to the FN and PLL domains grow following the pattern but show also a behaviour that is sensitive to both geometry and cell density. It was also noticed that cells on densely crowded islands stayed in non-differentiated mode longer than low populated samples where cell commitment to a differentiated mode is favoured. Figure 6 shows that by varying the conditions of culture (with and without serum, with and without growth factor, on FN and PLL patches), we found that the differentiation of HUCB-NSC could be modulated to a large extent, the population of neurons, astrocytes and oligodendrocytes varying between 10% and 80% depending on the conditions [9,16,62,63].

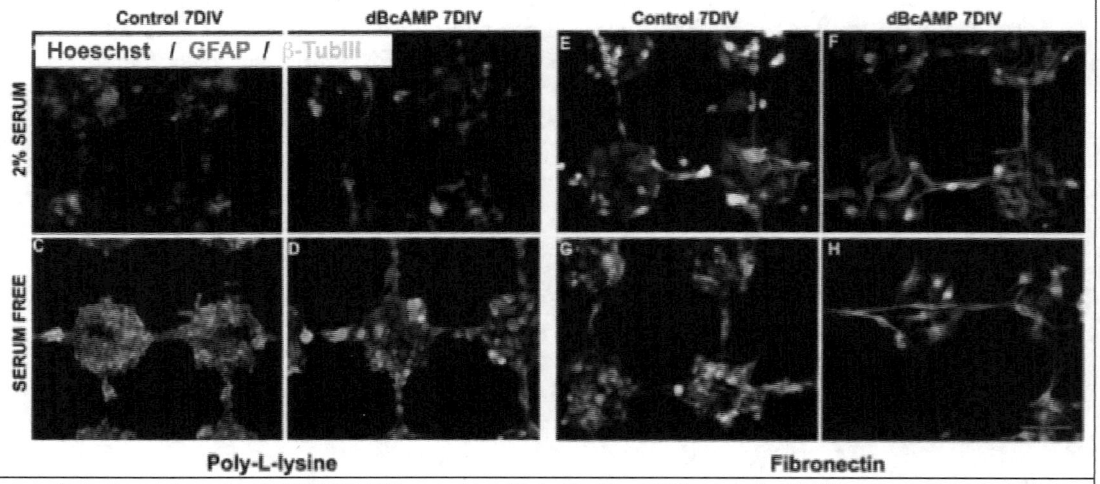

Figure 6: Immuno-images of HUCB-NSC grown on PLL (A-D) and fibronectin (E-H) microarrays for 7 days in serum (A, B, E, F) and serum free conditions (C, D, G, H) differentiating in the presence of dBcAMP (B, F, D, H) or growing as a control cultures (A, C, E, G). Neuronal (β-tubulin III) and astrocytic markers indicate the changes in differentiation depending on the PLL or FN markers. The scale bar is 100 μm for all images. Reproduced with permission from [9]

4. Conclusions

We have shown that plasma technologies, and in particular deposition, lithography and etching can be combined with colloidal and electron-beam lithography as well as microcontact printing for production of high quality micrometric and sub-micrometric patterned surfaces with arbitrary geometries and controlled chemistry. The major advantage of plasma polymers is related to the large range of properties that can be obtained, and their relative stability with time and against different solvents and on the fact that they can be applied on virtually any substrate. A major challenge of these processes will be optimization of the robustness and control of surface properties to make them applicable at a large scale at low cost.

5. References

[1] Walter G, Bussow K, Lueking A and Gloker J, 2002 *Trends. Mol. Med.* **8**, 250
[2] Zhu B, Zhang Q, Lu Q; Xu, Y; Yin J, Hu J andWang Z, 2004 *Biomaterials*, **25**, 4215
[3] Blawas A S and Reichert W M, 1998 *Biomaterials*, **19**, 595
[4] Curtis A and Riehle M 2001, *Phys. Med. Biol.* **46**, R47
[5] Arnold M, Cavalcanti-Adam A, Glass R, Blummel J, Eck W, Kantlehner M, Kessler H and Spatz J P 2004, *ChemPhysChem.* **5**, 383
[6] Thomas C H, Lhoest J B, Castner D G, McFarland C D and Heal K E, 1999 *Trans ASME J. Biomech. Eng.* **121** (1), 40
[7] Kam L, Shain W, Turner J N and Bizios R, 1999 *Biomaterials*, **20**, 2343
[8] Gadegaard N, Martines E, Iehle M O, Seunarine K and Wilkinson C D W, 2006 *Microelectronics Engineering*, **83**, 1577
[9] Bużańska L, Ruiz A, Zychowicz M, Rauscher H, Ceriotti L, Rossi F, Colpo P, Domańska-Janik K and Coecke S. 2009 *Acta Neurobiol. Exp.*, **69** 24-36
[10] Fink J, Théry M, Azioune A, Dupont R, Chatelain F, Bornensa M and Piel M, 2007 *Lab Chip*, **7**, 672–680
[11] Dalby M J, Berry C C, Riehle M O, Sutherland D S, Agheli H and Curtis A S G, 2004, *Experimental Cell Research* 295 387– 394

[12] Whitesides G M, 2006 *Nature* **442**, 368
[13] Seemann R, Brinkmann M, Kramer E J, Lange F L, and Lipowsky R, 2005, *PNAS* **102** (6) 1848-1852
[14] Beni V, Valsesia A, Colpo P, Brétagnol F, Rossi F and Arrigan D W M 2007 *Electrochemistry Communications,* **9**, 1833
[15] Valsesia A, Colpo P, Meziani T, Lisboa P, Lejeune M, and Rossi F, 2006 *Langmuir,* **22** (4), 1763
[16] Ruiz A, Buzanska L, Gilliland D, Rauscher H, Sirghi L, Sobanski T, Zychowicz M, Ceriotti L, Brétagnol F, Coecke S, Colpo P and F. Rossi, 2008 *Biomaterials* **29** 4766–4774
[17] Ploux L, Anselme K, Dirani A, Ponche A, Soppera O and Roucoules V 2009, *Langmuir*, **25**(14), 8161–8169
[18] Lercel M J, Craighead H G, Parokh A N, Seshadri K and Allara D L, 1996 *Appl. Phys. Lett.* **68**, 1504
[19] Lussi J W, Tang C, Kuenzi P A, Staufer U, Csucs G, Voros J, Danuser G, Hubbell J A and Textor M, 2005 *Nanotechnology*, **16**, 1781
[20] Haddour H, Cosnier S, and Gondran C, 2005 *Journal of American Chemical Society*, **127**, 5752-5753
[21] Renault J P, Bernard A, Bietsch A, Michel B, Bosshard H R, Delamarche E, Kreiter M, Hecht B and Wild U P J, 2003 *Phys. Chem. B.* **107**, 703
[22] Taha H, Marks R S, Gheber L A, Rousso I, Newman J, Sukenik C and Lewis A, *Appl. Phys. Lett.* **83**, 1041 (2003)
[23] Delamarche E, Schmid H, Michel B and Biebuyck H, 1997 *Advanced Materials*, b, 741.
[24] Rossini P, Colpo P, Ceccone G, Jandt K D and Rossi F, 2002, *Materials Science and Engineering C*, 1050, 1
[25] Manso M, Valsesia A, Ceccone G, Colpo P and Rossi F, 2004 *Appl. Nanoscience*, **1**(1), 47
[26] Lejeune M, Brétagnol F, Ceccone G, Colpo P and Rossi F, 2006 *Surface and Coatings Technology*, **200**, 5902
[27] Favia P, De Vietro N, Di Mundo R, Fracassi F and d'Agostino R, 2006 Plasma Process. Polym., **3**, 66–74
[28] Brétagnol F, Lejeune M, Papadopoulou A, Hasiwa H, Rauscher H, Ceccone G, Colpo P and Rossi F, 2006 *Acta Biomaterialia*, **2** (2), 165
[29] Chu L, Knoll W and Förch R, 2006 *Plasma Process. Polym.* **3**, 498–505
[30] Shyong Siow K, Britcher L, Kumar S, and Griesser H J 2006, *Plasma Process. Polym.* **3**, 392–418
[31] Sardella E, Favia P, Gristina R, Nardulli M and d'Agostino R, 2006 *Plasma Process. Polym.* **3** 456–469
[32] Lopez L C , Gristina R, Ceccone G, Rossi F, Favia P and d'Agostino R. 2005 *Surface & Coatings Technology* **200** 1000– 1004
[33] Valsesia A, Manso Silvan M, Ceccone G, Gilliland D, Colpo P and Rossi F 2005 *Plasma Process. and Polym.*, **2**, 334–339
[34] Valsesia A, Colpo P, Manso Silvan M, Meziani T, Ceccone G and Rossi F, 2004 *Nano Lett.* **4**(6); 1047
[35] Sardella E, Gristina R, Senesi G S, d'Agostino R and Pietro Favia P 2004 *Plasma Process. Polym.* **1**, 63–72
[36] Agheli H and Sutherland D S, 2006, *IEEE Trans. on Nanobioscience* **5** (1) 9-15
[37] Michel R, Reviakine I, Sutherland D, Fokas C, Csucs G, Danuser G, Spencer N D and Textor M, 2002, *Langmuir* 18, 8580-8586
[38] Dirani A, Wieder F, Roucoules V, Airoudj A and Sopera O, 2010, *Plasma Process. and Polym.* **8**, 571-581
[39] Chen W and Ahmed H, 1993 *Appl. Phys. Lett.* **62**, 1499
[40] Vieu C, Carcenac F, Pepin A, Chen Y, Mejias M, Lebib A, Manin-Ferlazzo L, Couraud L and Launois H, 2000 *Appl. Surf. Sci.* **164**, 111
[41] Brétagnol F, Ceriotti L, Lejeune M, Papadopoulou-Bouraoui A, Hasiwa M, Gilliland D, Ceccone G, Colpo P and Rossi F 2006, *Plasma Process. and Polym.* **3**, 30
[42] Baralia G G, Pallandre A, Nysten B and Jonas A M, 2006 *Nanotechnology*, **17**, 1160
[43] Detomaso L, Gristina R, Senesi G S, d'Agostino R and Favia P, 2005 *Biomaterials*, **26** (18), 3831
[44] Latour R., 2006 *Journal of Biomedical Materials Research* Part A Volume 78A, **4**, 843-854
[45] Ruiz S A and Chen C S 2007 *Soft Matter,* **3**, 1-11
[46] Kumar A and Whitesides G M 1993 *Appl. Phys. Lett.,* **63**, 2002-4
[47] Xia Y and Whitesides G M, 1998 *Angew. Chem. Int. Ed.* **37**, 550-75.
[48] Csucs G, Michel R, Lussi J W, Textor M and Danuser G 2003 *Biomaterials*, **24**, 1713-20
[49] Chang J C, Brewer G J and Wheeler B C, 2003 *Biomaterials*, **24**, 2863-70.
[50] Suh K Y,Seong J, Khademhosseini A, Laibinis P E and Langer R 2004 *Biomaterials*, **25**, 557-563.

[51] Ruiz A, Valsesia A, Brétagnol F, Colpo P, and Rossi F 2007 *Nanotechnology,* **18**, 531-536.
[52] Ruiz A, Buzanska L, Ceriotti L, Brétagnol F, Coecke S, Colpo P and Rossi F 2008 *J. Biomater. Sci. Polymer Edn.,* **19**(12):1649-57
[53] Delamarche, E. Donzel C, Kamounah F S, Wolf H, Geissler M, Stutz R, Schmidt-Winkel P, Michel B, Mathieu H J, and Schaumburg K, 2003 *Langmuir*, **19**, 8749-8758.
[54] Valsesia A, Colpo P, Meziani T, Brétagnol F, Lejeune M, Rossi F, Bouma A and Garcia-Parajo M, 2006 *Advanced Functional Materials*, **16**, 1246
[55] Valsesia A, Meziani T, Brétagnol F, Colpo P, Ceccone G and Rossi F, 2007 *J. Phys. D: Appl. Phys.* **40**, 2341
[56] Brétagnol F, Valsesia A, Sasaki T, Ceccone G, Colpo P and Rossi F 2007 *Advanced Materials*, **19** (15), 1947 (2007)
[57] Brétagnol F, Kylian O, Hasiwa M, Ceriotti L, Rauscher H, Ceccone G, Gilliland D, Colpo P and Rossi F, 2007 *Sensors and Actuators B*, **123**, 283
[58] Lisboa P, Valsesia A, Colpo P, Rossi F and Mascini M, 2010 *Analytical Letters*, **43**: 1556–1571
[59] Lisboa P, Valsesia A, Manelli I, Mornet S, Colpo P and Rossi F, 2008 *Adv. Mater.*, 9999, 1–7
[60] Valsesia A, Mannelli I, Colpo P, Bretagnol F and Rossi F, 2008, *Anal. Chem.* **80**, 7336–7340
[61] Valsesia A, Colpo P, Mannelli I, Mornet S, Bretagnol F, Ceccone G and Rossi F 2008 *Anal. Chem.* **80**, (5) 1418-1424
[62] Buzanska L, Jurga M, Stachowiak E K, Stachowiak M K and Domanska-Janik K 2006, *Stem Cells Dev.* **15**, 391-406.
[63] Buzanska L, Habich A, Jurga M, Sypecka J and Domanska-Janik K, 2005 *Toxicology in Vitro*, **19**, 991-999.

Honeycomb structured porous interfaces as templates for protein adhesion

Juan Rodríguez-Hernández,[1,2] Alexandra Muñoz-Bonilla,[1] Emmanuel Ibarboure,[2] Vanesa Bordegé,[1] Marta Fernández-García[1]

[1] Instituto de Ciencia y Tecnología de Polímeros (ICTP-CSIC), C/ Juan de la Cierva nº3 28006 Madrid Spain.
[2]Laboratoire de Chimie des Polymères Organiques, LCPO-ENSCPB-CNRS, 16, Av. Pey Berland 33600 Pessac (France)
E-mail: jrodriguez@ictp.csic.es

Abstract. We prepared breath figure patterns decorated with a statistical glycopolymer, (styrene-*co*-2-{[(D-glucosamin-2-N-yl)carbonyl]oxy}ethyl methacrylate, S-HEMAGl). The preparation of the glycopolymer occurs in one single step by using styrene and S-HEMAGl. Blends of this copolymer and high molecular weight polystyrene were spin coated from THF solutions leading to the formation of surfaces with both controlled functionality and topography. AFM studies revealed that both the composition of the blend and the relative humidity play a key role on the size and distribution of the pores at the interface. The porous films shows the hydrophilic glycomonomer units are oriented towards the pore interface since upon soft annealing in water, the holes are partially swelled. The self-organization of the glycopolymer within the pores was additionally confirmed both by reaction of carbohydrate hydroxyl groups with rhodamine-isocyanate and by means of the lectin binding test using Concanavalin A (Con A).

1. Introduction

Surface engineering of a biomaterial surface has to take into account several aspects including the modification and fine tuning of the surface chemical composition, interfacial morphology (roughness, topography) and optionally the coexistence and distribution of micro/macrodomains.[1] A large variety of recent examples have evidenced the key role of these parameters in the surface properties including adhesion/friction, wettability, or surface conductivity. For instance, the wettability can be varied between a hydrophilic and a superhydrophobic behavior by choosing the appropriate chemical function at the interface.[2] In particular, patterned surfaces serve as templates to create a pattern of proteins that can, in turn, be employed to study the interactions between proteins or antibodies with other biomolecules.[3]

Breath figures are formed by water condensation during solvent evaporation thus leading to a physical micropatterning ('top-down' approach) in which the topography of the polymer surface can be modified by creating holes of controlled dimensions. The affinity between the hydrophilic segments of the copolymer and the condensed water droplets, offers unique opportunities to modify the pore surface by with the glycomonomer units. The surfaces prepared following this approach could be of potential interest as 3D cell culture platforms. It is worth to mention that topographically controlled surfaces have been until now obtained by using tedious multistep procedures using lithographic techniques. In particular, glycopolymers, i.e. synthetic polymers containing carbohydrate units along or at the end of the main chain, because their "glyco-cluster effect", are interesting candidates to modify polymer surfaces creating bioactive materials. Carbohydrates are involved in multiple biological processes in living systems by participating in a variety of mutual recognition processes including immunological protection, virus infection, or recognition in the nervous system, amongst others.

We prepare regularly ordered 3D patterned polymer surfaces from blends of polystyrene and a statistical glycopolymer (styrene-*co*-2-(D-glucopyranosyl) aminocarbonyloxyethyl acrylate, S-HEAGl) using the breath figures methodology as dynamic templating method. In contrast to previous reported approaches in which the pores formed by using this methodology have been decorated with different functional groups or nanoparticles the preparation of breath figures with a glycopolymer is rather unusual. Nishikawa et al. in 1999 [4] described the preparation of poly-ion complexes at the surface with anionic polysaccharides. Very recently, Ting et al. [5] have reported the patterning of proteins onto galactosylated porous films. This article presents the formation of a regular film formation by using 6-arm star polystyrene which by itself present honeycomb structure with an ordered hexagonal array. Moreover, the preparation of the copolymer was carried out in multiple polymerization and protection/deprotection steps. The methodology employed throughout this manuscript leads to glycopolymers in one single step without protective chemistry. In addition, the parameters (humidity, temperature, polymer concentration, etc.) that significantly modify the surface topography in the breath figures mechanism have been widely investigated. Finally, we demonstrate the capability of the surfaces to selectively interact with proteins within the cavities.

2. Materials and methods

2.1. Materials

Styrene (St), (Aldrich, 99%) was distilled under reduced pressure. 2,2'-azoisobutyronitrile (AIBN) (Aldrich, 98%) was purified by successive crystallizations from methanol. The synthesis of 2-{[(D-glucosamin-2-N-yl) carbonyl]oxy}ethyl methacrylate (HEMAGl) will be described elsewhere. All the solvents used in the course of experiments and characterization (dimethyl formamide (DMF) and dimethyl sulfoxide-d) were employed without further purification. Rhodamine B isothiocyanate (Aldrich) and lectin-fluorescein isothiocyanate conjugate from Canavalia ensiformis (ConA-FTIC) (Sigma) were used as received.

2.2. Copolymerization reaction

Free-radical copolymerization of styrene and HEMAGl with a molar feed composition in styrene of 0.80 was carried out in pyrex ampoule sealed in argon atmosphere at 70°C in a global concentration of 1 mol/L DMF solution with $3 \cdot 10^{-2}$ mol/L of AIBN as the initiator concentration. The reaction was conducted up to total conversion. The resulting copolymer, S8Gl2, was purified by dialysis and further lyophilization.

2.3. Characterization

The statistical glycopolymer was characterized by ^{1}H-NMR with a Bruker Advanced 400 MHz spectrometer at room temperature. A probe of deuterated DMSO was introduced in a copolymer solution of dimethylformamide (DMF). Average molecular weights and dispersities were determined by size exclusion chromatography (SEC) using a Jasco system equipped with two PL gel 5 µm (300 x 7.5 mm) mixed-C columns and a PL gel 5 µm (50 x 7.5 mm) guard column, a Jasco 1530 differential refractive index detector and a Jasco 875 UV detector. N,N-Dimethylformamide (HPLC grade) was used as eluent containing 0.1 M LiBr with a flow rate of 0.8 ml/min at 60 °C. Calibration was obtained from narrowly distributed polystyrene standards.

2.4. Film preparation

Mixtures having 10-50 wt% of copolymer S8Gl2 and 90-50% high molecular weight polystyrene (M_n= 250000 g/mol) were dissolved in THF (solution concentration 30 mg/mL). The polymer solutions were filtered with a 0.1 μm Millipore membrane and spin-coated (4000 rpm for 60 s) onto silicon wafers purchased from SC (Siegert Consulting e.K., Germany). The silicon wafers were cleaned prior to use in piranha solution (3:1 v/v of H_2SO_4 in H_2O_2) and rinsed several times with ethanol. For the preparation of samples under controlled relative humidity, beakers containing water and saturated aqueous solutions of sodium bromide were placed inside the spin coating chamber in order to obtain values of relative humidity between ~40 and ~57 % respectively. The humidity degree and the temperature were measured by means of a hygrothermograph.

2.5. Annealing

The sample annealing was carried out to further study the variations of the surface chemical composition as a function of the environment. Upon analyzing (by AFM imaging) the films obtained after spin-coating, the samples were exposed either to air at 90 °C for 3 days or placed in a tightly closed stainless steel vessel saturated in water vapor. The annealing in a humid environment was carried out during 36 h at 90 °C. After each treatment, the samples were dried under vacuum at room temperature.

2.6. Rhodamine-isocyanate immobilization

The films previously annealed to water vapor were immersed in an aqueous Rhodamine-isocyanate solution (0.25 mg/mL) for 2 h at room temperature. After washing with deionized water, the film was dried at room temperature.

2.7. Lectin interaction

The film was immersed into a phosphate-buffered saline solution (PBS, 7.4) at room temperature containing 0.2 mg/mL of fluorescein-conjugated Concanavalin A (Con A-FITC). After 6 hours, the film was sequentially washed with PBS buffer and distilled water.

2.8. Methods

Atomic Force Microscopy, AFM, images were recorded in air at room temperature with a Nanoscope IIIa microscope operating in tapping-mode. The probes were commercially available silicon wafer with a spring constant of 42 N/m, a resonance frequency of 285 KHz and a typical radius curvature in the 10-12 nm range. Both the topography and the phase signal images were recorded with a resolution of 512x512 data points.

Fluorescence Microscopy and Image Processing. Images were performed on a Zeiss Axiovert 40 CFL inverted microscope equipped with a 12-V, 35-W halogen lamp (for the phase-contrast images) and an HBO 50 W/AC mercury lamp (for the fluorescence images). The objectives used were a 5×/0.12 A-Plan, a 10×/0.25 A-Plan, a 20×/0.50 EC Plan-NEOFLUAR, and a 40×/0.50 LD A-Plan (Zeiss). Images were acquired by using a CANON A640 CCD camera.

3. Results and discussion

The preparation of structured interfaces was carried out by mixing a statistical copolymer (Mn= 42900 g/mol; PD=1.73) prepared by conventional free radical polymerization with linear high molecular weight polystyrene (Mn=250000 g/mol). The copolymer consists of styrene and 2-{[(D-glucosamin-2-N-yl) carbonyl]oxy}ethyl methacrylate (HEMAGl) units in 80 to 20 molar ratio. The composition of the copolymer can be varied by modifying the initial feed. Herein, we employed a copolymer with a relative high amount of styrene (80% of styrene and 20% of HEMAGl). This proportion was chosen due to the following reasons: first, by increasing the amount of styrene content, the compatibility between the glycopolymer and the homopolymer matrix is enhanced. Second, since in a subsequent step the prepared films will be annealed in water. The amount of hydrophilic glycomonomer has to be low in order to form stable films that do not dissolve in aqueous media and could resist the annealing conditions.

Films of the blends varying the relative composition of glycopolymer to linear polystyrene matrix were prepared by spin coating from THF solutions (30 mg/mL) in an atmosphere under controlled humidity (44% RH). As it has been reported, during solvent evaporation the temperature of the solvent-air interface decreases and water vapor starts to condense (Figure 1). As the solvent evaporation continues the condensed water droplets grow until, finally, upon complete evaporation of both solvent and water the resulting surface contains holes randomly distributed reflecting the positions where the water droplets condensed. The conditions in which the films were prepared influence the size and distribution of the holes obtained.[6] Hence, solvent, blend composition, concentration of solution and humidity are important external parameters that have to be controlled.

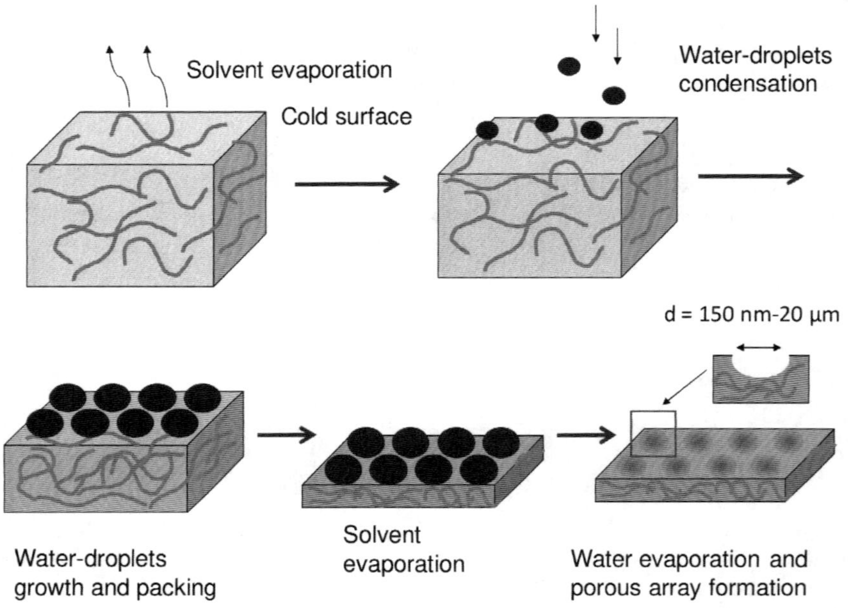

Figure 1. Breath figures approach to obtain porous polymeric interfaces by water condensation.

In Figure 2 are illustrated the AFM images of films obtained for a blend composed of 20% in weight of statistical glycopolymer and 80% of polystyrene. Amphiphilic copolymers are hygroscopic and may interact with water during the spin coating process. Thus, increasing the amount of hydrophilic polymer in the mixture enhances water uptake and as a consequence the diameter of the holes formed. Blends with amounts of S8Gl2 above 50% lead to rather disordered surface structures as a consequence of the coalescence of the water droplets. Equally, films prepared in highly moist atmospheres above 50% RH lead to a similar effect. Hence, within this range of blend compositions and humidity the average size of the holes can be controlled.

Journal of Physics: Conference Series **252** (2010) 012002 doi:10.1088/1742-6596/252/1/012002

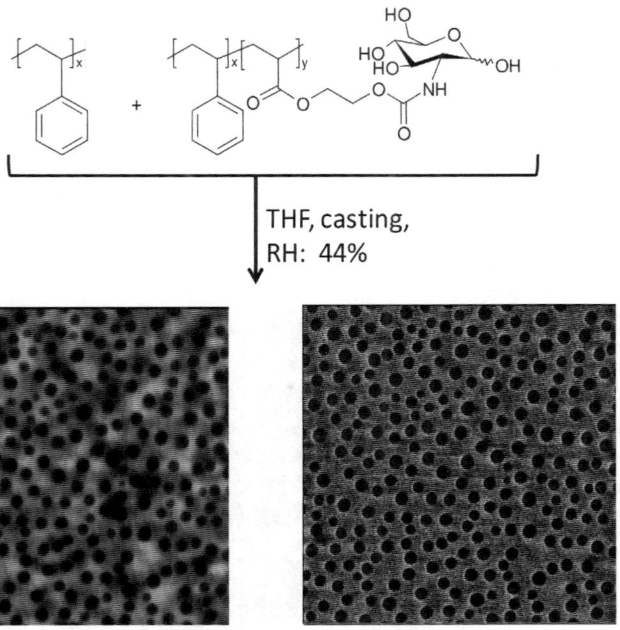

Figure 2. AFM height (left) and phase images (right) (10µm x10µm) of films obtained by spin coating in a 44% humid atmosphere with a blend composition of 20 /80 (w/w, S8Gl2/PS).

The interaction between the polar glycomonomer units within the copolymer and water droplets condensed during the spin coating modifies the distribution in terms of chemical composition at the interface. As a consequence, the glycopolymer in contact with water rearranges around the water droplet and after drying is principally localized in this area. Hence, the surface composition is different between the holes and the rest of the interface. The hygroscopic nature of the carbohydrate moieties reveals upon annealing in humid air a preferential swelling this area. This behavior is evidenced by AFM in films obtained directly after spin coating and annealed in humid air. Annealing in humid water vapor provokes reversible changes on the micropatterned surface. Hence, the holes are transformed in island-like structures after 3 days of annealing. Similarly, drying the surface under mild conditions (60°C, overnight) removes the water and recovers the original topography.

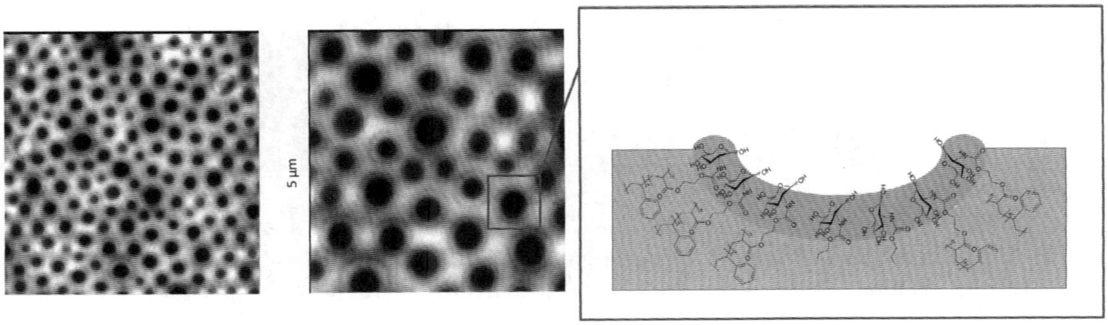

Figure 3. Tapping mode height AFM images (10 µmx10µm and 5µm x 5µm) of: film obtained by spin coating 50/50 S8Gl2/PS (w/w) after annealing in water vapor at 80 °C for 3 days.

The employment of the surfaces as templates needs the availability of recognition/reactive sites to which the biomolecules could be attached. Whereas the presence of glycopolymer and therefore of saccharide units within the pore surface is proven by AFM upon imaging the surfaces obtained after annealing in water vapor, the availability of the hydroxyl groups of the saccharide units is evidenced by reaction with rhodamine isothiocyanate (Rho-FITC). The images obtained with a fluorescence microscope for the blends with 50% of S8Gl2 are depicted in Figure 4. The ring-shaped fluorescent pattern indicated the attachment of the rhodamine to the pore surface by covalent reaction between the isothiocyanate and the hydroxyl moieties. These images clearly indicate the capability of the hydroxyl groups to accomplish chemical reactions with other functional groups.

The ability to these glycopolymer-modified surfaces to specifically recognize lectins is evidenced by using Concanavalin A which is a protein which specifically interacts with glucopyranosyl residues. The films are immersed into PBS solution at room temperature containing 0.2 mg/mL of fluorescein-conjugated Concanavalin A (Con A-FITC). After 6 hours, the film is sequentially washed with PBS buffer and distilled water. Figure 4 (right) shows the fluorescence microscopy of the interaction between the S8Gl2 glycopolymer with Con A-FITC. As can be observed, such interaction is produced in the external part of the pore confirming that the hydrophilic part of glycopolymer is mainly exposed to the surface.

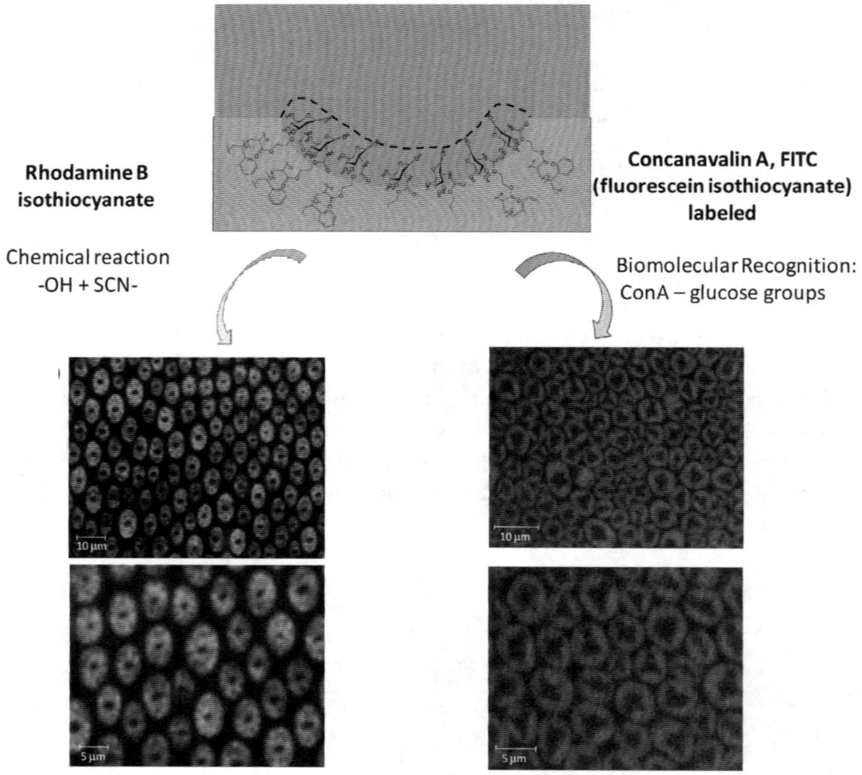

Figure 4. Top view images of the films obtained by using a fluorescence microscope. Left: the fluorescence detected in the inner part of the pore evidenced the functionalization of the inside of the pore with a rhodamine-isocyanate by reaction with the hydroxyl groups of the glycomonomer units. Right: images of the surface obtained upon chemical recognitition between the glycopolymer and fluorescein-conjugated Concanavalin A within the pores.

4. Conclusions

We described the elaboration of micropatterned surfaces by using the breath figures methodology. By controlling the parameters involved in the film preparation such as, composition of the blend, humidity and temperature we are able to prepare surfaces with defined pore size and distributions. By swelling the polar carbohydrate moieties we evidenced the selective enrichment of glycopolymer in the inner part of the pore. Depending on the size of the pores created, the pores are either transformed in hills or only partially swollen. Finally, the potential of these structures to serve as templates for the attachment of bioactive molecules is clearly evidenced. Firstly, the hydroxyl groups of the saccharide are able to react inferring specific functionality as demonstrated by the reaction with the isothiocyanate functional group of the rhodamine. And secondly, the specific interaction with Con A lectin clearly verified the accessibility of the carbohydrate groups into the surface, indicating again the transformation from hydrophobic to hydrophilic surface.

5. Acknowledgments

The authors gratefully acknowledge financial support from the Spanish National Research Council (CSIC) through the PI 200860I037, MAT2009-12251, MAT2007-60983 and CCG08-CSIC/MAT-3643 programs.

References

[1] (a) Tsukruk VV, 1997 *Prog. Polym. Sci.* **22** 247. (b) Lafuma A, Quere D, 2003 *Nat.Mater.* **2** 457. (c) Graighead H.G. 2000 *Science* **290** 1532.

[2] (a) Bormashenko E, Pogreb R, Stanevsky O, Bormashenko Ye, Socol Y, Gendelman O, 2005 *Polym. Adv. Technol.* **16** 299. (b) Nishikawa T, Nishida J, Ookura R, Nishimura Sh.-I, Wada Sh, Karino T, Shimomura M, 1999 *Mater. Sci. Eng. C* **10** 141.

[3] (a) Dusseiller MR, Smith ML, Vogel V, Textor M, 2006 *Biointerphases* **1** P1. (b) Dusseiller MR, Schlaepfer D, Koch M, Kroschewski R, Textor M, 2005 *Biomaterials* **26** 5917.

[4] Nishikawa T, Nishida J, Ookura R, Nishimura S.-I, Wada S, Karino T, Shimomura M, 1999 *Mater. Sci. Eng., C* **8-9** 495.

[5] Ting SRS, Min EH, Escale P, Save M, Billon L, Stenzel M H, 2009 *Macromolecules* **42** 9422.

[6] Bunz UHF 2006 *Adv. Mater.* **18** 973.

Design and rapid prototyping of DLC coated fractal surfaces for tissue engineering applications

A Díaz-Lantada[1], A. Mosquera[2], J L Endrino[2], P Lafont[1]
[1] Machines Engineering Research Group, Universidad Politécnica de Madrid
c/ José Gutiérrez Abascal 2, 28006 Madrid, Spain
[2] Surfaces and Coatings Department, Instituto de Ciencia de Materiales de Madrid
c/ Sor Juana Inés de la Cruz 3, Cantoblanco, 28049 Madrid, Spain

E-mail: adiaz@etsii.upm.es

Abstract. Several medical devices (both implantable and for in vitro diagnosis) benefit greatly from having microtextured surfaces that help to improve and promote phenomena such as osteointegracion and cell / tissue growth on the surface of a device. Normally, the use of abrasives or chemical attacks are employed for obtaining such surface microtextures , however, it is sometimes difficult to precisely control the final surface characteristics (porosity, roughness, among others) and consequently the related biological aspects.
In this work, we propose an alternative process based on the use of fractal surface models for designing special surfaces, which helps controlling the desired contact properties (from the design stage) in multiple applications within biomedical engineering, especially regarding tissue engineering tasks. Manufacturing can be directly accomplished by means of rapid prototyping technologies. This method supposes a focus change from a conventional "top-down" to a more versatile "bottom-up" approach.
Finally, in order to improve the possible biological response, the surfaces of the designed devices were coated with hydrogen-free amorphous carbon (a-C) thin films, known to be highly biocompatible materials. The films were deposited at room temperature using the vacuum filter cathodic arc technique. Our first prototypes have helped verify the viability of the approach and to validate the design, manufacturing and coating processes.

Key words. Fractal, scaffold, tissue engineering, rapid prototyping, diamond-like carbon.

1. Introduction: Fractals in biomedical engineering

Fractals are rough or fragmented geometric shapes that can be split into parts, each of which is (at least approximately) a reduced-size copy of the whole. The term *fractal* was coined by Benoît Mandelbrot in 1975 [1] and derives from the Latin *fractus* meaning "broken" or "fractured". The term "fractal" is used to describe complex geometries that are too complex to be described in conventional Euclidean terms, with properties like self-similarity and defined usually with simple recursive procedures.

[1] To whom any correspondence should be addressed.

Since the early works linked to fractal geometry in 1975, it became clear that they could be used for describing the geometries, patterns and roughness of natural objects [1-2]. Although fractals are commonly considered to be infinitely complex (due to their usual recursive definitions) "approximate fractals" are easily found in nature, which usually display self-similar structure over an extended, but finite, scale. By limiting the steps applied in a recursive definition of a conventional fractal, approximate fractals can be obtained, which mimic complex natural geometries. Natural objects that are approximated by fractals include clouds, mountains, lightning bolts, coastlines, snow flakes, various vegetables and several corporal geometries.

During the last decade, increasing attention has been paid to using fractals for promoting modeling, design and simulation tasks in several areas of Bioengineering. The most remarkable are discussed below:

- Modelling the behaviour of microorganisms.

  Several studies have been reported on the use of fractal models for describing the growth and expansion rate of bacteria and for evaluating the dynamics of coexisting species of microorganisms [3].
- Modelling complex organisms and their systems.

  Regarding complex organisms (including human anatomy) fractals have been applied to modelling systems of pulmonary and blood vessels and vascular networks, as well as for carrying out subsequent fluid mechanics simulations [4].
- Modelling the surfaces of organs and tissues.

  Recent interest has appeared in the use of fractals for mimicking the surfaces of organs and tissues and thus improving the designs and *in vivo* performance of several prosthetic devices [5].

In this study, we focus on the design, manufacture and biological conditioning through a surface coating of fractal surfaces for obtaining special rapid-manufactured scaffolds with controlled features such as roughness, waviness, skewness or fractal dimension, among others.

2. Benefits of fractal scaffolds and current limitations

Several studies have focused on the importance of surface microtexture, roughness and pore size for promoting positive effects in all kinds of biodevices, from implantable prosthesis to scaffolds for cell and tissue growth. These textures have a significant influence in osseointegration of prosthesis, cell proliferation and tissue growth given that those cells and tissues seem to be more "comfortable" when faced with biodevices with similar surface properties [5-6].

However, the process of introducing desired roughness on the surfaces of biodevices is still mainly linked to carrying out machining operations, laser processing or chemical attacks. In all these cases, post-processing operations could be difficult to control. The use of fractal models for mimicking such natural surfaces can prove to be useful for design tasks. The process explained in the following section allows defining and controlling the texture and roughness of surfaces from the design stage, with help of computer-aided design tools. Its application to obtaining fractal scaffolds is shown as example.

The mentioned computer-aided design, calculation and manufacturing technologies (CAD-CAE-CAM), have become essential tools for developing products. They allow 3D geometries and alternative designs (to which calculations of stress, deformations, ergonomics, dynamic response can be applied) to be rapidly manufactured. Moreover, these technologies can be employed to test materials and designs before manufacturing. The multiple benefits of these technologies for developing conventional products could prove to be extremely beneficial when applied to designing biomimetic and personalized scaffolds for tissue engineering, although some limitations exist.

Some enterprises already provide their customers with scaffolds, designed with the help of CAD tools, however they are limited to conventional design operations from such software (holes, grooves, extrusions) and the related scaffolds do not provide the desired complexity.

Additionally the use of CAD tools can be especially beneficial when combined with a new set of manufacturing techniques and technologies that have appeared in the last two decades called "Rapid prototyping and manufacturing technologies". These technologies help to address market requirements in an ever more customised way and to provide support for research work where physical models (or prototypes) are needed for tests and trials that are optimised in terms of time and cost.

They are usually based on manufacturing processes that add layers or "layer manufacturing technologies" (like "laser stereolithography", "3D printers" or "selective laser sintering"), in rapid shape-copying processes, or in manufacturing processes through the elimination of material (such as in high speed numerical control machining). The different technologies available allow prototypes to be obtained rapidly in a wide range of metallic, ceramic or polymeric materials with remarkable precision [7-8].

These technologies enable physical parts to be manufactured in a short time (a couple of hours or a few days) directly from the designs made with the help of computer-aided design, calculation and manufacturing programs or "CAD-CAE-CAM technologies". They greatly help to optimise design iterations, contribute to early error detection and speed up production start-up.

3. Designs of fractal scaffolds

We propose and explain in this section the use of mathematical fractal models for designing the surfaces of the scaffolds and the use of these surfaces, together with CAD programs, for obtaining three-dimensional virtual prototypes of the scaffolds for subsequent automatic manufacture. In this way, parameters such as roughness, waviness, skewness can be controlled from the design stage and adapted in a more efficient way to the requirements of the damaged organ or tissue.

For our initial designs we have selected a fractional Brownian fractal surface model, which has previously proved to be useful when carrying out designs of natural surfaces [2]. The following equation gives the height "z" of the surfaces, when assessing the function over a grid of points given by their (x,y) coordinates. The model uses several random functions (A_k, B_k, C_k), several control constants (λ, α, m) and an initial height function "z_0" can also be introduced.

$$z(x,y) = z_0(x,y) + m\cdot\sum_{k=1}^{\infty} C_k \cdot \lambda^{-\alpha k} \cdot \sin(\lambda^k [x \cdot \cos(B_k) + y \cdot \sin(B_k) + A_k]) \quad (1)$$

Figure 1 shows the result of evaluating a fractional Brownian fractal function over a grid of 60 x 60 points (corresponding to a scaffold of 30 mm x 30 mm) and the influence of introducing changes in the control parameter "α". Figure 1 (a) corresponds to α = 0.8 (fractal dimension around 2.2 with maximum roughness depth reaching 1.2 mm) and Figure 1 (b) corresponds to α = 0.2 (fractal dimension around 2.8 with maximum roughness depth reaching 2.5 mm). In these examples we have used $z_0(x,y) = 0$, what leads to positive and negative values of z in the images of Figure 1, as the z-axis shows. The calculations have been carried out with help of Matlab software (Mathworks, version R2009) and the data obtained are stored in three-column matrixes [X, Y, Z]. The command "surf" helps to represent the surfaces linked to the mentioned matrixes.

Once the Matlab surfaces have been obtained, the information stored in the form [X, Y, Z] can be converted into .stl universal format, so that the surface can be recognized and imported with a CAD program, for additional design operations (i.e. providing the surface with a thickness different than zero, copying the surface atop from a previously designed geometry). The explained process is currently patent pending, both for two-dimensional fractal surfaces with application in tissue engineering [9] and for three-dimensional fractal structures with application in biofabrication [10].

The rapid manufacture of such fractal scaffolds and the use of DLC coating for promoting their biocompatibility are described in the following section.

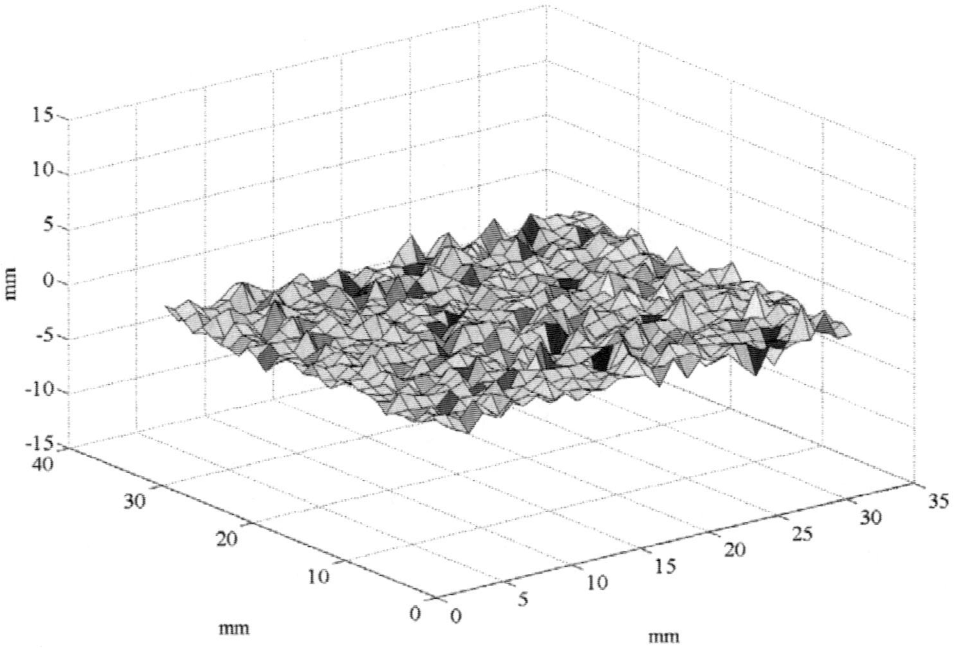

(a)

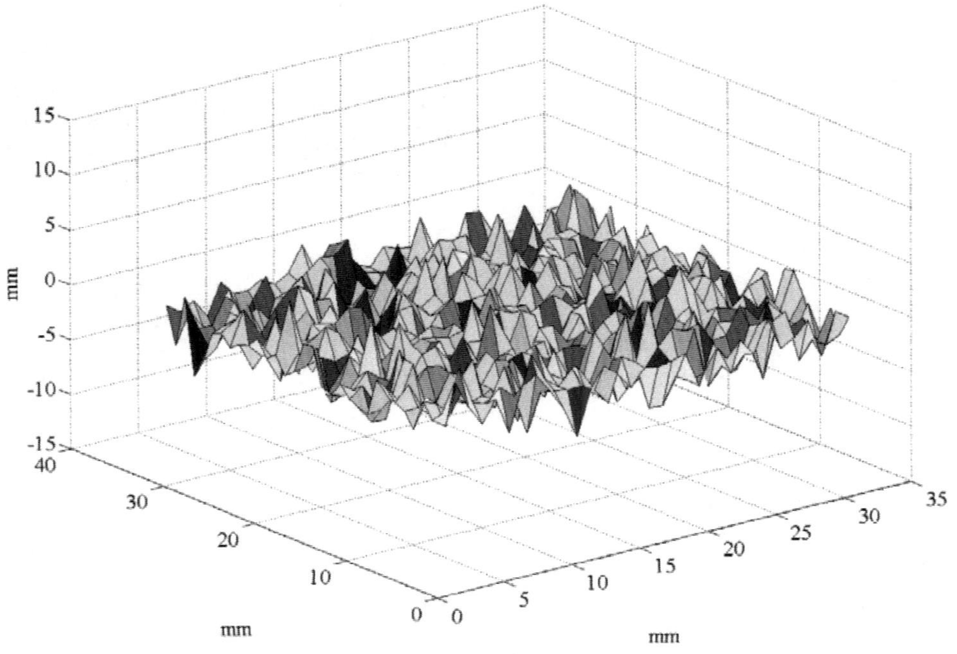

(b)

Figure 1. Designs of fractal surfaces with different roughness and fractal dimension.(a) $\alpha = 0.8$, fractal dimension around 2.2 and (b) $\alpha = 0.2$, fractal dimension around 2.8.

4. Manufacture of fractal scaffolds

In spite of the fast incorporation of rapid prototyping into product development methodologies, these new types of technologies are still in their initial stages of development, hence their applications in bioengineering (and especially in tissue engineering) are continuously evolving [11-12].

In this study we propose the use of a rapid prototyping technology (laser stereolithography) for directly manufacturing fractal scaffolds from the previously explained designs. Some enterprises already provide their customers with rapid-prototyped scaffolds, however they are normally designed using conventional CAD design operations (holes, grooves, extrusions) and the related prototypes do not provide the desired complexity. The design and manufacturing approach proposed here provides additional control tools and helps to obtain biomimetic scaffold designs, which can be directly manufactured. Figure 2 shows an example of a physical prototype of fractal scaffold manufactured through laser stereolithography in epoxy resin (Accura 60, 3D Systems), directly from the CAD file of the part. It has been manufactured using the SLA-3500 machine from (3D Systems) which can manufacture details down to around 150 μm.

It is important to note that using a layer manufacturing technology allows to obtain very complex geometries, even with inner details and porosity, in many cases unable to be manufactured using conventional top-down approaches such as machining or chemical attack. Aspects like roughness, waviness, skewness or fractal dimension can be precisely controlled from the design stage and adapted in a personalized way to the requirements of the application.

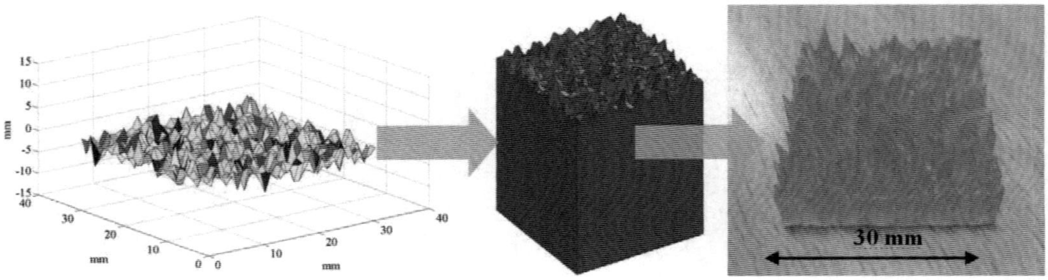

(a) (b) (c)

Figure 2. Fractal surface (a), CAD model derived from fractal surface (b) and prototype of fractal scaffold (c). Physical prototype obtained through laser stereolithography in epoxy resin.

Although the photopolimerizable epoxy resin used for the first trials in not a medical grade polymer, and therefore cannot be used for the manufacture of implantable devices, it is important to mention some recent advances in the development of biocompatible photopolimerizable resins for stereolithography, which will help in the following years to overcome this limitation [13].

In the meantime we propose two possibilities for tackling the problem of biocompatibility. Using the epoxy fractal scaffold we can obtain silicone moulds by rapid form copying for subsequent casting of medical grade polymers conventionally used for scaffolds (polyvinyl alcohol, polycaprolactone, polydimethylsiloxane). For *in vitro* diagnosis and cell culture tasks, a more direct procedure through the deposition of a highly-biocompatible coating can be a more appropriate solution.

5. Diamond-like carbon coatings for property enhancement

After obtaining the prototypes, hydrogen-free amorphous carbon thin films (a type of diamond-like carbon or DLC with sp3 content below 30%) were deposited at room temperature using the vacuum filter cathodic arc technique [14].

Remarkable adhesion and film conformity were obtained in the initial trials; however some small fissures appeared in some of the sharper tips of the prototype due to the high aspect-ratio of some of the features, as shown in Figure 3, and the induced stress produced by the deposition of the a-C film. However, tips fissure problems can be avoided by the use of scalonated and smoother designs, as shown in Figure 4. DLC coatings with thickness of around 100 nm were successfully applied using the cathodic arc technique.

In any case, these initial trials helped to show the viability of successfully coating rapid-prototyped epoxy parts with diamond-like carbon coatings, which are highly biocompatible as reported in previous work [15]. Moreover the possibility of combining the benefits of amorphous carbon with some quantities of metal (especially Ag or Cu) as doping agents can lead to surfaces with enhanced anti-bacterial properties, especially useful when taking into account the possibility of implanting such devices [16-17].

It is important to note that the coating process has to be carried out at room temperature, which allows coating polymers, with glass transition temperatures around 40 – 60 ºC, without inducing remarkable deformations or morphological changes. Thus, final devices, even after coating, possess the geometrical properties defined from the initial fractal model used and afterwards converted into the CAD file.

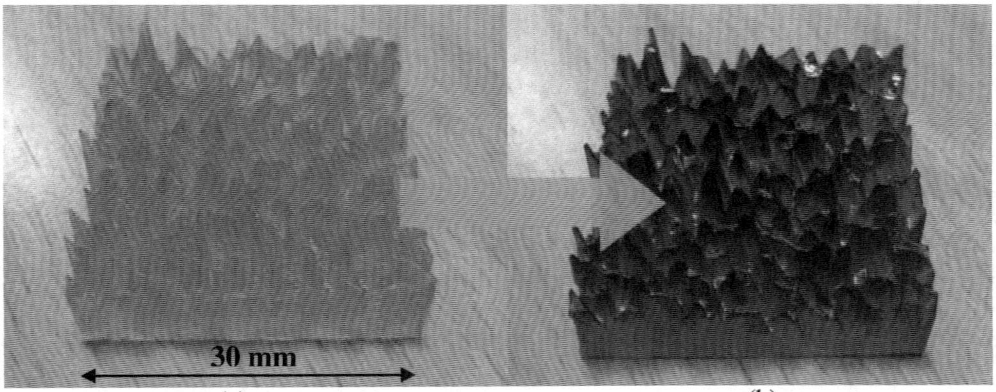

(a) (b)

Figure 3. Pictures of fractal scaffold prototype: (a) uncoated, (b) after DLC coating.

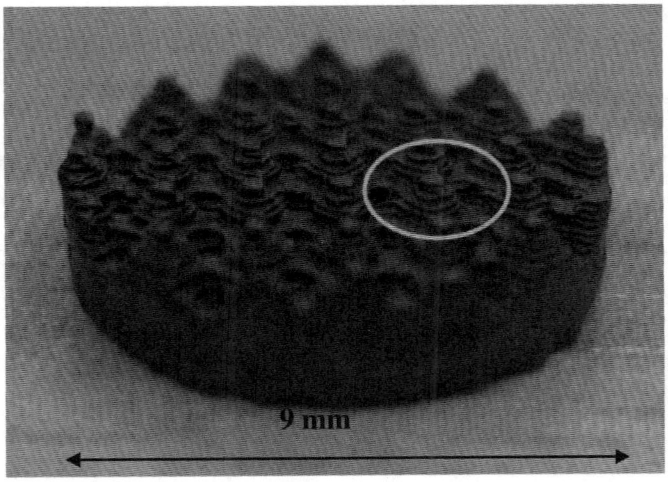

Figure 4. Detail of a second prototype of rapid manufactured scaffold with DLC coating. Layer by layer process can be perceived in detail inside the image.

6. Conclusions and future work

We have presented a novel method for designing, manufacturing and enhancing biocompatibility of scaffolds for tissue engineering based on the use of fractal surfaces for promoting biomimetic approaches. Several designs and prototypes have been obtained for validating the different stages of the process.

Linking fractal models with CAD programs and rapid manufacturing technologies proves to be a powerful tool for precisely controlling (from the design stage) the final properties of the scaffolds like roughness, waviness, skewness or fractal dimension.

Main future actions are aimed at *in vitro* validation of the prototypes obtained and at the verification of the coating influence on biocompatibility. Furthermore, it is important to explore the use of microstereolithography for increasing the accuracy of our prototypes and for enabling the application of the proposed methodology to solve additional bioengineering needs.

Acknowledgements

Authors would like to thank Prof. Jose Maria Albella Martín for his helpful comments. We acknowledge the support of the Spanish Ministry of Science and Innovation through the Singular Strategic Project "IBE–Rapid Manufacturing" (PSE-020000-2009-1), Consolider Project CSD 2008-0023 FUNCOAT and Project FIS2009-12964-C05-04. Additionally, Universidad Politécnica de Madrid has supported the idea of ***using fractal geometries for tissue engineering tasks*** (www.fractal-bio.com) through its "Actua UPM Innovation and Spin-off Creation Programme" (www.upm.es/actuaupm) & (http://actuaupm.blogspot.com).

References

[1] Mandelbrot B B 1982 The Fractal Geometry of Nature (W.H. Freeman and Company)

[2] Falconer K 2003 *Fractal Geometry: Mathematical Foundations and Applications* (John Wiley & Sons, Ltd.)

[3] Tsyganov M A, Kresteva I B, Aslanidi G V, Aslanidi K B, Deev A R and Ivanitsky G R 1999 The Mechanism of Fractal-Like Structure Formation by Bacterial Populations *Journal of Biological Physics* **25** 2 165-176

[4] Lin D W, Johnson S and Hunt C A 2004 Modeling Liver Physiology: Combining Fractals, Imaging and Animation *Proceedings of the 26th Annual International Conference of the IEEE EMBS* San Francisco CA USA September 1-5 3120-3123

[5] Longoni S and Sartori M 2010 Fractal geometry of nature (bone) may inspire medical devices shape *Nature Proceedings* (in press) received 02 June 2010 posted 03 June 2010

[6] Choi J W, Wicker R, Lee S H, Choi K H, Ha C S, Chung I 2009 Fabrication of 3D biocompatible / biodegradable micro-scaffolds using dynamic mask projection microstereolithography *Journal of Materials Processing Technology* **209** 5494-5503

[7] Wohlers T 2010 *Wohlers' Report: Additive Manufacturing State of the Industry* (Wohlers Associattes)

[8] Kucklick T R 2006 *The Medical Device R&D Handbook* (CRC Taylor & Francis)

[9] Díaz Lantada A, Lafont P, Albella J M, Endrino J L, Ortego P, Muñoz J L, Muñoz J M, Echávarri J 2010 Substrato cuasibidimensional para crecimiento de células y tejidos y método de obtención del mismo *Spanish Patent and Trademark Office* Patent application number **P201030957**

[10] Díaz Lantada A, Lafont P, Albella J M, Endrino J L, Ortego P, Muñoz J L, Muñoz J M, Echávarri J 2010 Soporte tridimensional para producción artificial de órganos y otras estructuras del organismo humano y método de obtención del mismo *Spanish Patent and Trademark Office* Patent application number **P201030956**

[11] Díaz Lantada A, Lafont P, Lorenzo H, Muñoz J, Muñoz J L, Echavarri J, Muñoz J M 2009 Rapid prototyping and rapid tooling technologies for developing shape-memory polymer based devices *IV ECCOMAS Thematic Conference, Smart Structures and Materials (SMART'09)*

[12] Díaz Lantada A, Del Valle Fernández R, Lafont Morgado P, Muñoz García J, Muñoz Sanz J L, Munoz J M, Echávarri J 2010 Development of Personalized Annuloplasty Rings: Combination of CT Images and CAD-CAM Tools *Annals of Biomedical Engineering* **38** 2 280-290

[13] Stampfl J, Schuster M, Baudis S, Lichtenegger H, Liska R, Turecek C and Varga F 2007 Biodegradable stereolithography resins with defined mechanical properties *Virtual and Rapid Manufacturing (ed: P.J. Bartolo et al.), Proceedings VRAP 2007* 283-288

[14] Zhang H, Endrino J L, and Anders, A 2008 Comparative surface and nano-tribological characteristics of nanocomposite diamond-like carbon thin films doped by silver *Applied Surface Science* **255** 5 Part 2 2551-2556,

[15] Salgueiredo E, Vila M, Silva M A, Lopes M A, Santos J P, Costa F M, Silva R F, Gome P S and Fernandes M H 2008 Biocompatibility evaluation of DLCcoated Si3N4 substrates for biomedical applications *Diamond Related Materials* **17**(4–5) 878–881

[16] Endrino J L, Horwat D, Gago R, Andersson J, Liu Y S, Guo J, et al. 2009 Electronic structure and conductivity of nanocomposite metal (Au, Ag, Cu, Mo)-containing amorphous carbon films *Solid State Sciences* **11** 10 1742-1746 ,

[17] Endrino J L, Escobar Galindo R, Zhang H, Allen M, Gago R, Espinosa A, et al. 2008 Structure and properties of silver-containing a-C(H) films deposited by plasma immersion ion implantation *Surface and Coatings Technology* **202** 15 3675-3682

Computer Simulation of Scaffold Degradation

G Erkizia[1], A Rainer[2], E M de Juan-Pardo[1] and J Aldazabal[1]

[1] CEIT and Tecnun (University of Navarra), Manuel Lardizábal 15, 20018 San Sebastián, Spain

[2] CIR - Laboratory of Chemistry & Biomaterials, University Campus Bio-Medico of Rome, Via Alvaro del Portillo 21, Rome, Italy

E-mail: gerkizia@ceit.es

Abstract. Scaffolds are porous biocompatible materials with suitable microarchitectures that are designed to allow for cell adhesion, growth and proliferation. They are used in combination with cells in regenerative medicine to promote tissue regeneration by means of a controlled deposition of natural extracellular matrix by the hosted cells therein. This healing process is in many cases accompanied by scaffold degradation up to its total disappearance when the scaffold is made of a biodegradable material.

This work presents a computational model that simulates the degradation of scaffolds. The model works with three-dimensional microstructures, which have been previously discretised into small cubic homogeneous elements, called voxels. The model simulates the evolution of the degradation of the scaffold using a Monte Carlo algorithm, which takes into account the curvature of the surface of the fibres.

The simulation results obtained in this study are in good agreement with empirical degradation measurements performed by mass loss on scaffolds after exposure to an etching alkaline solution.

1. Introduction

In recent years, biomedical research has been developing strategies to regenerate tissues in the search for new treatments for damaged or diseased organs. Some of these strategies, included in the so-called regenerative medicine field, are the replacement, repair and/or regeneration of tissues (e.g. bone, cartilage, heart valves, bladder, etc.). These tissues substitutes need to fulfil certain requirements on the mechanical and structural properties in order to work properly.

Within the field of regenerative medicine, Tissue Engineering applies engineering principles to produce biological substitutes that maintain, improve or restore the function of organs or tissues of the human body [1]. New biomaterials are often used to replace the damaged part of an organ that has ceased in its functions, or in other cases, as vehicles to transport cells and molecules to the target tissue or organ. These new tissular substitutes act as "scaffolds" that allow for cell growth and formation of natural extracellular matrix.

For a correct operation, the artificially designed scaffolds must be biocompatible, should not present any adverse immunological reaction and in most cases need to be gradually degraded as new biological tissue is formed, thus transferring the withstand of the mechanical loads progressively to the newly formed autologous tissue. The main objective of this work is to model the degradation of scaffolds by hydrolysis. Such degradation will take place in the gradual replacement of a scaffold during bone regeneration.

The time required from the design of a biomaterial scaffold to its clinical application is very long, since it requires several validation steps, thus increasing the overall costs of the final product. It has been estimated that the average development period for a tissue engineering product can range up to 10 years. Therefore, substantial investments are required, in terms of laboratory equipment, men power and validation tests, until the commercialization of the product is reached.

In this framework, it is clear how *in silico* experiments, based on accurate computer models, could help reduce the overall costs of final tissue engineering products in a drastic way.

2. Model

The model proposed in this work was coded in C and was executed in a Linux workstation (DELL Precision T5400). Following subsections show the ideas and principles behind each part of the model.

2.1. Initial microstructure

The model works on scaffolds consisting of equally spaced parallel layers of fibres. The initial microstructure is defined analytically by several parameters: volume size, number of layers, fibres diameter, spacing between fibres in the same layer and tilt angle of the fibres on each layer. The model assumes that the distance between two consecutive layers is zero, i.e., consecutive layers touch each other physically.

2.2. Volume discretisation

Once the microstructure is defined, it has to be discretised into cubic elements. It is assumed that the total volume is composed of many identical cubes, called voxels. Each of these voxels may contain information about the corresponding phase -matrix or fibre- and, in the latter case, to which fibre it belongs. For each voxel, the concentration of the component fibre can also be determined.

The accuracy of the discretisation process can be adjusted by controlling the size of voxels. Certainly, the number of voxels used for discretising a given volume strongly affects computational effort and requirements.

Working with cubic elements has the following advantages [2]:

- All voxels are of equal size.
- The number of neighbours for all voxels is constant, as well as the area shared with each surrounding neighbour.
- For a voxel, the position of the vertices can be derived from the coordinates of its centre.
- Coding benefits due to the simple geometry of a voxel.

2.3. Boundary conditions

Periodical boundary conditions can not be systematically applied as a rule to a microstructure. The geometry of the scaffold will depend on the number of fibres per layer and the tilt angle. Once these parameters are fixed, it is frequent that the final microstructure is not periodic. As a consequence of this lack of periodicity, a special set of boundary conditions must be applied, where the simulated volume should be extended over a voxel on each side, that is, if the domain size along the i axis is l_i the model should work with a volume with a dimension of $l_i + 2$ voxels. In this way, the so called extended volume is obtained to define the boundary conditions of the simulation. The extended faces parallel to fibre layers are considered in the model as liquid phase, i.e., there are no fibres, while on the other four sides the extended volume is completely filled with “fibres voxels”.

Figure 1 represents the geometric relationship between the simulated volume and the extended volume. The simulated volume is constituted by different fibre layers, as shown in the first image (a). Parallel to these layers, the extended volume is filled by the liquid phase (b) and perpendicular to the layers (c) the extended volume is filled with a solid phase, i.e. with fibre voxels.

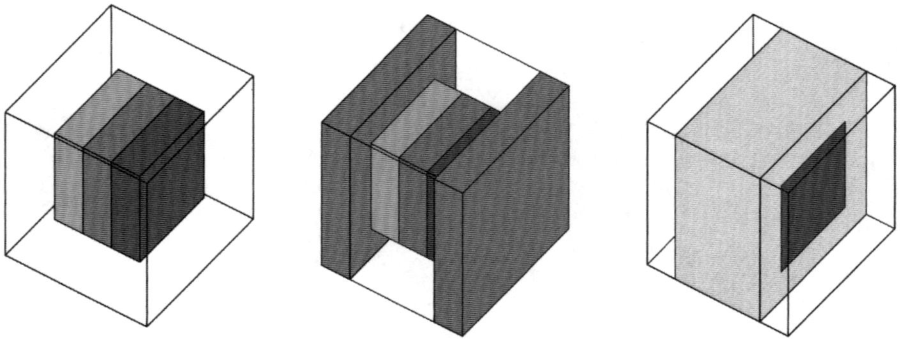

(a) (b) (c)

Figure 1. Construction of the extended volume around the simulated volume. Different boundary conditions for the extended volume are reported (b,c).

2.4. Monte Carlo Method

In this work we have used the Monte Carlo Q-Potts algorithm to simulate the degradation of scaffolds [3].

Once the initial three-dimensional microstructure has been discretised, the Monte Carlo method randomly selects a voxel within the simulated volume (not within the extended volume). The dissolution probability of a voxel of the scaffold is related with the status (solid or liquid) of its surrounding neighbours. The algorithm works by counting how many neighbours of each phase are present. For example, if the chosen voxel is solid, i.e., belongs to a fibre, and has many liquid neighbours, its dissolution probability is high. If the voxel belonging to the solid phase is inside the fibre, i.e. all its neighbours are solid, it will never dissolve. This process is repeated for all voxels in the simulated volume.

In this kind of models it is very common to measure the magnitude of the simulation time using as a time unit Monte Carlo Steps (MCS). A MCS is reached when all voxels of the domain have been chosen randomly once, i.e. the whole volume was sampled. One MCS is equivalent to a number of trials equal to the total number of voxels.

2.5. Assigning weights to neighbours

There are, in total, three types of neighbours: face, edge or vertex neighbours. Since the influence of each type of neighbour is different, different weights should be assigned to each one. The proposed model considers only two types of neighbours: face and edge neighbours. The weight assigned to each type of neighbour is inversely proportional to the distance between the voxel and the neighbour. These weights are shown in Table 1 [4].

2.6. Assigned probabilities

Before defining the probability distribution function (PDF), some considerations are needed.

When a voxel is completely surrounded by solid neighbours, i.e. there is no surrounding liquid, it cannot be dissolved, so its dissolution probability is 0.

Table 1. Normalised neighbour weights

Neighbour type	Number of neighbours	Weight	Normalised weight
Face	6	1	0.069
Edge	12	$1/\sqrt{2}$	0.049
Maximum weight		$6(1+\sqrt{2})$	1.000

$$P_d(0) = 0 \quad (1)$$

On the other hand, when a solid voxel is in the middle of the liquid, it will be always dissolved.

$$P_d(1) = 1 \quad (2)$$

Summarising, we can state that the probability of dissolution, P_d, depends on the number of liquid neighbours, n. Between the two extreme cases previously described, there are other key configurations to determine the PDF.

In Figure 2a, the model tries to dissolve the voxel at the surface. This configuration has a normalised number of liquid neighbours n equal to 0.26, therefore it will sometimes be allowed to initialize the degradation of the flat surfaces.

Figure 2b shows another key case. This configuration corresponds to a thin edge with 0.60 normalized neighbours. The probability of dissolution assigned to this case is equal to one: that is, if a voxel is chosen randomly within an edge like this will always dissolve.

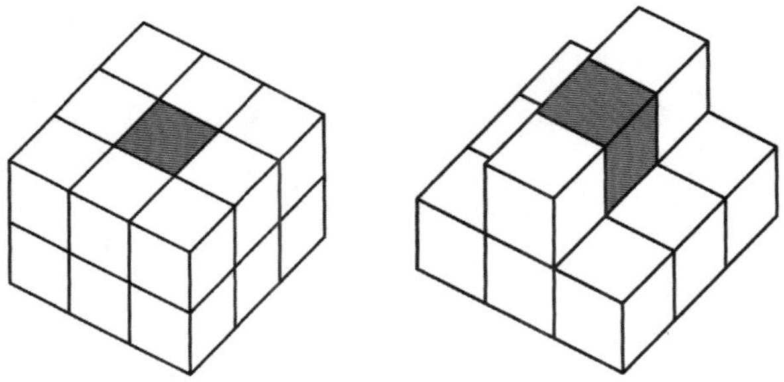

(a) (b)

Figure 2. Key cases to generate the PDF for voxels Degradation.

Several tests were performed to ensure an isotropic behaviour. This behaviour was obtained assigning a probability of 0.1 to the configuration shown in Figure 2a [5].

$$P_d(0.26) = 0.1 \quad (3)$$

Once these key cases were studied a linear probability distribution between these two cases was assigned, as shown in Figure 3.

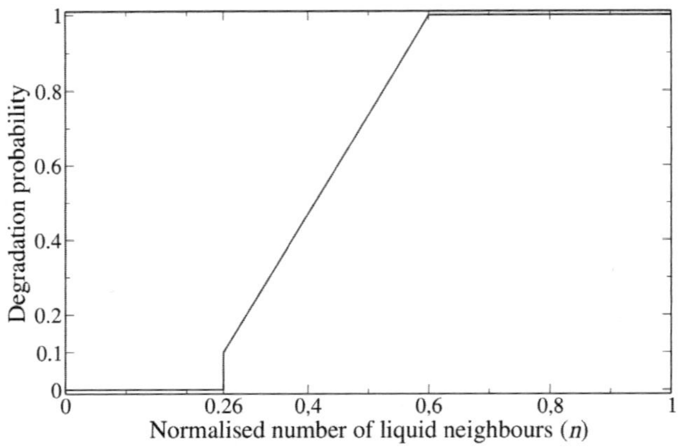

Figure 3. Probability distribution function used for degradation based on the normalized number of liquid neighbours.

2.7. Degradation

The degradation simulated in this model is hydrolysis, which is related to the contact surface between fibres and liquid.

As mentioned above, the method used to degrade the fibres in the microstructure is based on the Monte Carlo algorithm.

In this method, the standard number of liquid neighbours is measured. As shown in Figure 3, it is possible to know the probability of degradation of a selected voxel. The number of liquid neighbours for a specific voxel will determine its probability of degradation, using a randomly generated number between 0 and 1 as a comparator. If the probability to dissolve is greater than the comparator, the voxel will be degraded, i.e. converted into liquid; and if the probability is lower, the voxel will remain in its previous configuration, i.e. as a solid fibre.

3. Results

To validate the model, results of the simulations were compared with empirical degradation experiments of a scaffold made of poly-ε-catrolactone/β-tricalcium phosphate (PCL/TCP). PCL is a semi-crystalline and biodegradable polymer belonging to the family of aliphatic polyesters, while TCP is a biocompatible and bioactive ceramic material with a stoichiometry similar to the mineral phase of bone. It is widely used as a scaffold in bone tissue engineering due to its osteconductivity and its ability to bond directly to bone. The empirical data were obtained from the literature, relative to a scaffold degradation in a diluted NaOH solution, a process commonly used to accelerate the scaffold degradation by hydrolysis in in-vitro experiments [6].

The simulated geometry of the scaffold was generated modelling the real microstructure of the scaffold used in the empirical tests performed in [6], as shown in Figure 4.

The model was adjusted to generate a scaffold of a fibre diameter of 266 μm and a scale of 20 μm/voxel, while the cube side was adjusted to a simulation volume 4000 μm. 15 multifibre-layers were arranged with a tilt angle of 90°, 30°and -30°, and distance between fibres belonging to the same layer of 866 μm.

Once the microstructure was generated, it was introduced into the model to simulate its degradation. Figures 5 and 7 show the evolution of the geometry obtained by the model during the degradation. From these results, it is possible to confirm that the degradation process obtained with the model is homogeneous.

In particular, Figure 5 shows a comparison between the empirical data obtained on real

Figure 4. SEM micrograph of a real scaffold [6], 2D model projection and 3D simulated microstructure generated.

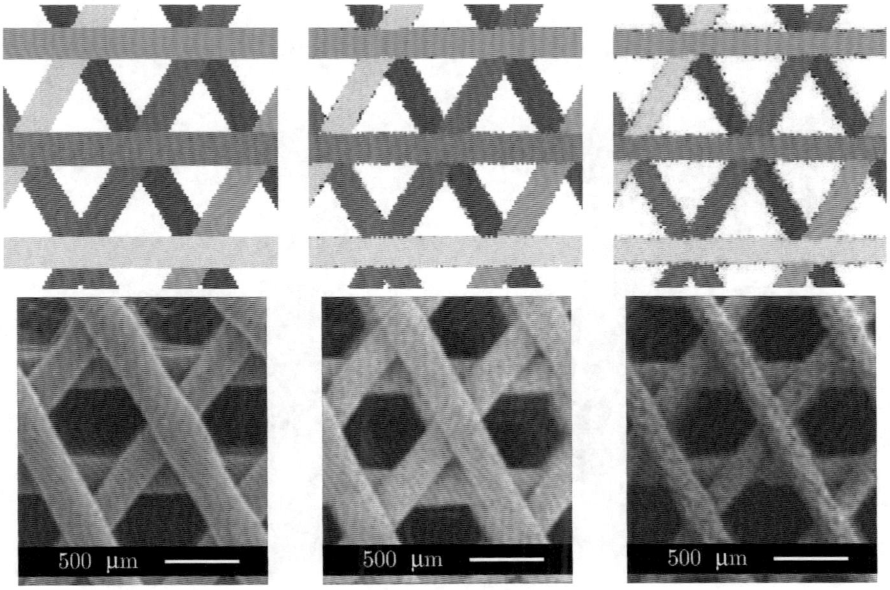

Figure 5. Comparison between model and real measurements [6].

scaffolds (SEM micrographs) and the results of the simulation process, in the 0-48 hours time range. Results confirm that the simulated degradation matches very well the empirical degradation data.

Given the correctness of the qualitative behaviour of the model, we tried to quantify the degradation process by estimating the evolution of the volume fraction of the solid phase. In order to easily compare computational results with empirical data, a conversion rate from simulated to real-time was estimated: 1 MCS $\approx$ 4.16 hours.

Figure 6 shows the empirically measured values for volume fraction versus the results of the computational model.

From Figure 6, it is possible to conclude that the model adjusts very well to empirical data.

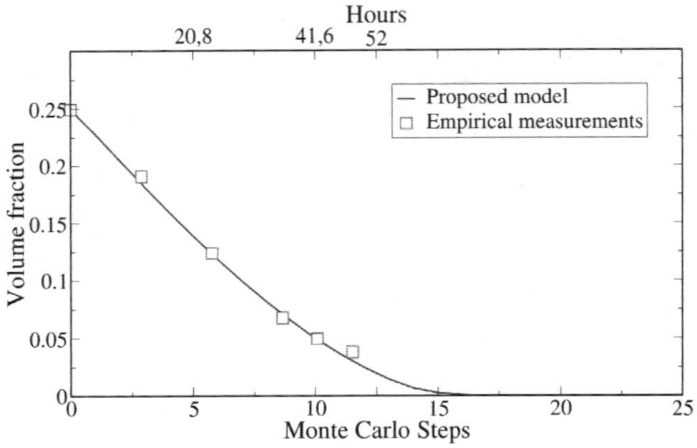

Figure 6. Evolution of volume fraction of fibres: comparison of the results obtained by the computational model (dividing the number of voxels belonging to fibres by the total number of voxels) versus the empirical mass loss measurements obtained by Lam et al in [6].

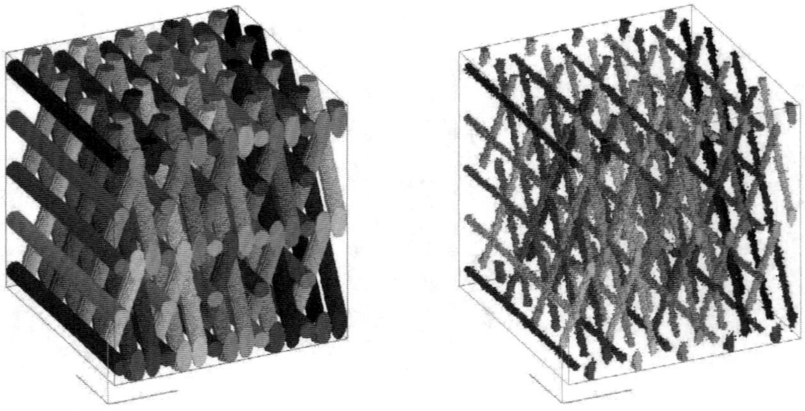

(a) (b)

Figure 7. Initial and final steps of the 3D simulation.

4. Conclusions

The proposed model is capable of generating solid geometries which resemble real scaffold microstructures.

The proposed degradation algorithm is isotropic and reasonably matches the empirical results for the degradation of scaffolds under accelerated hydrolysis tests. Yet, validation of the model against other real geometries needs to be performed.

References

[1] Mao J J, Vunjak-Navokovic G, Mikos A G and Atala A 2007 *Translational Approaches in tissue engineering and regenerative medicine* (Artech House Publishers)
[2] Rosenblum L 1994 *Scientific Visualization: Advances and Challenges* (Academic Press)
[3] Raabe D 1998 *Computational materials science: simulation of materials, microstructures and properties* (Wiley-VCH)
[4] Aldazabal J, Martin-Meizoso A and Martinez-Esnaola J M 2004 *Mater. Sci. and Eng.* **A365** 151-5

[5] Luque A, Aldazabal J, Martin-Meizoso A, Martinez-Esnaola J M, Gil Sevillano J and Farr R S 2005 *Modelling Simul. Mater. Sci. and Eng.* **13** 1057-70
[6] Lam C X F, Savalani M, Teoh S and Hutmacher D W 2008 *Biomed. Mater.* **034 108**, p.15

Beneficial silver: antibacterial nanocomposite Ag-DLC coating to reduce osteolysis of orthopaedic implants

J. L. Endrino[1*], J.C. Sánchez-López[2], R. Escobar Galindo[1], D. Horwat[3], A. Anders[4]
[1]Instituto de Ciencia de Materiales de Madrid (CSIC), Madrid, Spain
[2]Instituto de Ciencia de Materiales de Sevilla (CSIC-US), Seville, Spain.
[3]Institute Jean Lamour, Ecole des Mines de Nancy Nancy, France
[4]Lawrence Berkeley National Laboratory, Berkeley, California

*E-mail: jlendrino@icmm.csic.es

Abstract. Silver-containing diamond-like-carbon (DLC) is a promising material for biomedical implants due to its excellent combination of antibacterial and mechanical properties. In this work, a dual-cathode pulsed filtered cathodic arc source containing silver and graphite rods was employed in order to obtain DLC samples with various silver contents. Chemical composition of the samples was analyzed by acquiring their compositional depth-profiles using radio-frequency Glow Discharge Optical Emission Spectroscopy (rf-GDOES), while the microstructural properties were analyzed by X-ray diffraction and Raman spectroscopy. Tribological studies carried out against UHMWPE balls in fetal bovine serum indicate that the presence of silver in DLC could be beneficial to reduce the wear of the polymeric surfaces.

Keywords. Silver, amorphous carbon, Raman, GDOES, wear, biotribology.

1. Introduction

Diamond-like carbon (DLC) films are of interest to many applications due to their favorable mechanical and tribological properties [1]. They have also generated interest in the biomedical field because of their high biocompatibility and perform well as a surface that will support appropriate cellular activity [2-6]. Strictly speaking, the term “DLC” includes a broad range of metastable amorphous carbon materials with variable sp^2/sp^3 and H contents. In this manuscript, we will employ the broad term “DLC” to refer to hydrogen-free amorphous carbon with an overall sp^3 content around 30%, and deposited by the plasma immersion implantation and deposition technique. Amorphous carbon films can be functionalized using plasma treatments [7] and by embedding metallic elements, which can also add functionality by providing an antimicrobial effect [8-13]. However the use of DLC-coated implants has not been widely employed due to the increased wear observed in ultra-high molecular weight polyethylene (UHMWPE) in polymer/coated metal implants. The formation of worn particles of UHMWPE has been found to cause a medical condition called osteolysis, which refers to bone resorption due to removal or loss of calcium from the bone near the joint implant [14]. In this work, we carry out a set of tribological tests in order to determine the influence of silver embedded in a DLC matrix on the reduction of the wear of the UMWPE balls.

2. Experimental

Amorphous carbon films doped with silver were prepared on TiAl6V4 disc substrates by pulsed dual-cathode filtered cathodic vacuum arc (PDC-FCVA). Pulsed arc discharges on individual cathodes were triggered by a computer control system using a "triggerless" system [15]. The pulses used in the production of carbon plasma were 100 A and had a duration of 5 ms, while the metal plasma was made using pulses of 1 ms long and a current of 700 A. The plasma stream produced by the source is injected into a 90-degree filter to remove most of the macroparticles which were formed during the cathodic arc process. After exiting the filter, the plasma expanded through a homogenizer device composed of concentric magnets and allowed to have homogeneous thickness and composition in the substrates over an area of 80 cm^2. During the deposition of carbon pulses, the substrates were biased with negative 1 kV pulses that were 2 μs long with an off time between pulsed of 14 μs. The substrate was kept at ground potential during the deposition of the metal plasma (hence "species selective bias") [16]. The pulse repetition rate was set at 2 pulses per second and the number of pulses per deposition was 5000. During the deposition, the substrate holder was rotated at a speed of 2 rpm. The residual gas pressure was typically in the 10^{-4} Pa range. The thicknesses of the deposited films ranged from 50 nm to 100 nm.

Raman spectroscopy was used to characterize the microstructure of the carbon films and the changes caused by the silver addition. Raman spectra were acquired with a Jobin Yvon HR 460 monochromator and a nitrogen-cooled CCD (charge coupled device) detector. The excitation light was the 514.5 nm line of an Ar–Kr laser. The incident and scattered beams were focused using an Olympus microscope, and a Kaiser Super-Notch filter was used to suppress the elastically scattered light. The structural characteristics of the Ag-DLC films were probed by Grazing Incidence X-ray Diffraction (GIXRD) at an incident angle of 2° with respect to the substrate surface. The experiments were done with a Co Kα source. Glow Discharge Optical Emission Spectroscopy (GDOES) depth-profile analysis of the deposited films was done using a Jobin Yvon RF GD Profiler equipped with a 4 mm diameter anode. The GDOES was operating at a typical radio-frequency discharge pressure of 650 Pa and power of 40 W. Quantified profiles were obtained automatically using the standard Jobin Yvon QUANTUM Intelligent Quantification software.

The tribological properties were studied by reciprocating motion following the guidelines of the standard test method ASTM F732 for evaluating the friction and wear of materials for use in total joint prosthesis. UHMWPE 10-mm balls were rubbed against coated TiAl6V4 specimens in diluted fetal bovine serum (proteins content 20-23 g/l) in a CSM tribometer. Preliminary trials were attempted using flat-ender circular cylinders or conical shape tips but the difficulties to estimate the worn volume caused by different surface finishing made advisable the employment of balls to ensure similar contact conditions. The test parameters were set at 10 N of applied load (average contact pressure 30 MPa); 50 mm/s of linear speed; frequency 1 Hz and 75000 cycles (3.5 km of sliding distance). The tests were repeated at least twice to check the reproducibility. The worn surfaces were examined by using a Scanning Electron Microscope (SEM) JEOL JSM-5400 at 0.5 kV. The specific ball wear rate was estimated after dividing the worn cap volume by the applied load and the sliding distance.

3. Results

3.1 Composition and structural properties

Fig. 1 shows the GDOES compositional profile of silver for samples obtained with C:Ag pulse ratios of 24:1, 12:1, 6:1 and 3:1 (pulses of carbon:pulses of silver). There is a monotonous increase of the Ag content when decreasing the C:Ag ratio. The overall amount of silver was below 5 at.% for all composite samples, except in the case of 3:1 where the overall amount of silver was around 15 at.% and contained a large peak at the surface where the silver was concentration reached ~55 at.%. The segregation of certain amounts of silver to the surface is an expected effect due to the lower surface energy of silver compared to DLC. Fig. 2 shows the XRD patterns of the Ag-containing DLC samples. Crystalline silver peaks were only observed in samples containing 4 at.% and 15 at.%. of silver. These results are consistent with the formation of silver nanocrystals inside an amorphous carbon matrix.

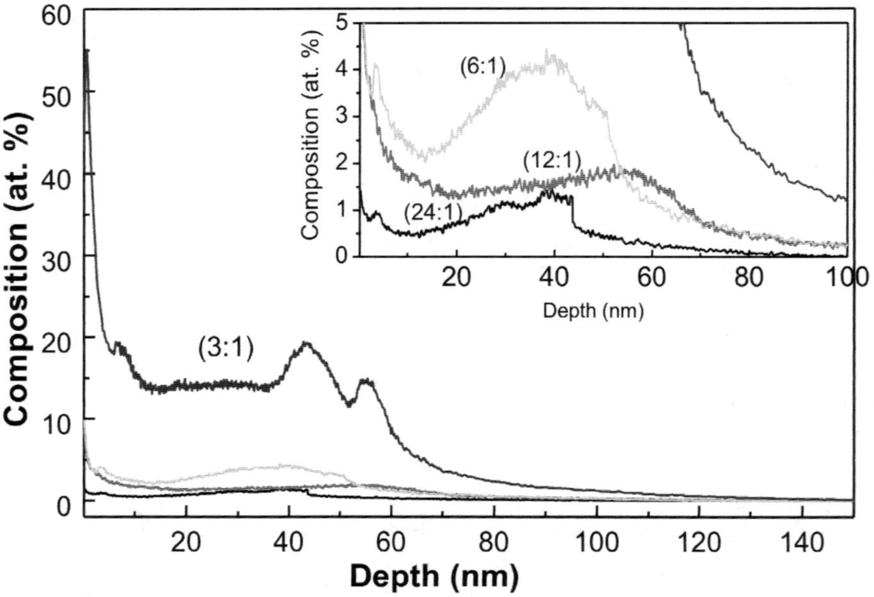

Figure 1. GDOES profile of silver in Ag-DLC coatings deposited by dual-cathode filtered cathodic arc.

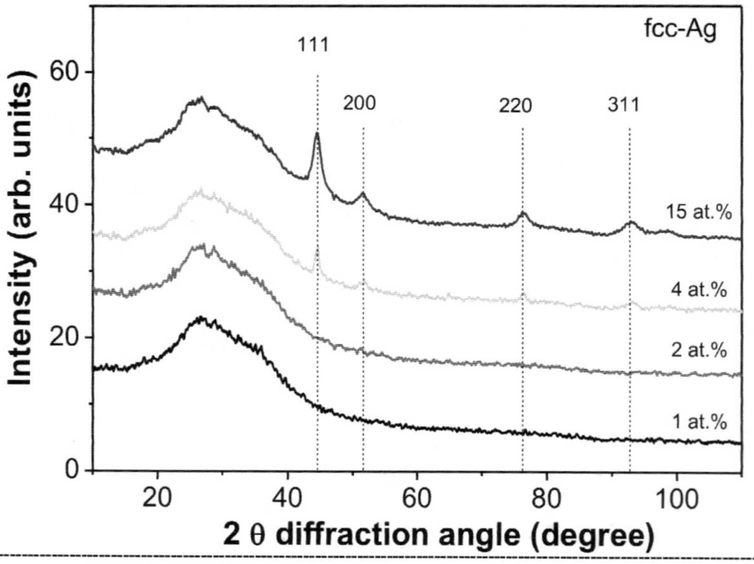

Figure 2. X-ray diffraction patterns of Ag-DLC coatings with various silver contents.

Normalized Raman spectra for the mixed Ag-DLC samples along a pure a-C sample are shown in Fig. 3. The Raman spectra for all the samples are featureless and have just one asymmetric broad peak with a maximum at around 1530 cm^{-1}. The only remarkable difference is the decrease of the total intensity of the peak with increasing the Ag content in the coating. The spectra which has been fitted before using two Gaussians centered at frequencies around 1350 cm^{-1} and 1530 cm^{-1}, commonly named as the D (disorder) and G (graphite) bands [17, 18], clearly shows that the peak position, full widths at half maximum (FWHM) and the ratio between the intensities of D and G peaks does not change significantly and implies that the amount of sp^3 and structural order in the amorphous carbon matrix remains intact and independent of the silver content. This is in good accordance with the formation of nanocomposite Ag-DLC.

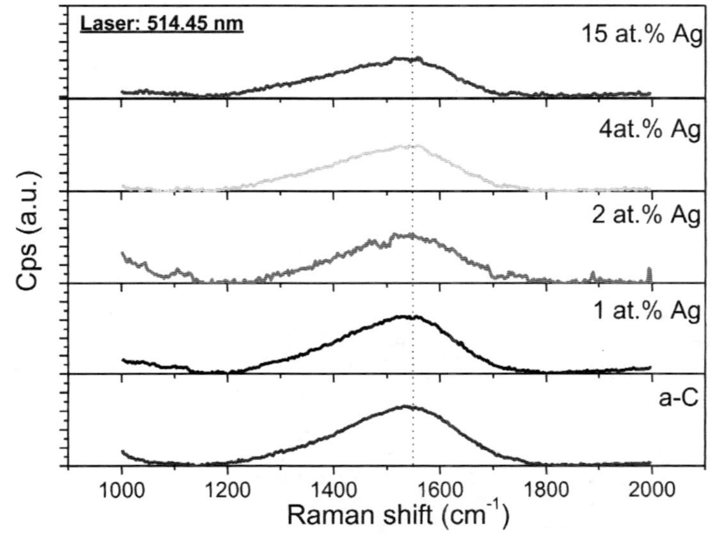

Figure 3. Raman spectra of Ag-DLC coatings collected at 514.45nm wavelength

3.2 Tribological Properties

The friction coefficients (shown in Fig. 4) of the Ag-DLC coatings varied between 0.07 and 0.09 being higher than the pure silver film whose friction value was about 0.04. They did not exhibit a clear dependence on the Ag content although an optimum was found for a small silver addition (2 at. %). Regarding the ball wear rate values, although the differences were not very significant (5 to 6.5×10^{-7} mm^3/Nm), a decreasing trend is observed upon incorporating silver in the carbon coatings.

Fig. 5 shows the optical micrographs taken from the ball counterfaces for the undoped carbon and with Ag (15 at.%) samples at the end of the friction tests. The rough surfaces of the UHMWPE balls appear smoothed in the contact region. The presence of silver seems to decrease the wear damage as can be denoted from the picture taken of the Ag (15 at.%)-DLC coating. The wear mechanism seems to be abrasive in all the cases identified by the presence of parallel groves onto the surface along the sliding direction. Under compression, the PE ball suffers plastic deformation and removal by the harder counterface. The slight diminution of the wear can be explained by a softening of the coatings when increasing the Ag content as reported by other authors [19,20].

Nevertheless, in order to determine whether the presence of silver on the top surface could have a tribochemical effect, a SEM analysis was carried out on the scar obtained with the pure Ag film. Fig. 6a illustrates the smoothing of the ridges originally present on the surface of the PE ball along the sliding direction. An EDX analysis carried out in this area (not shown) did not find evidence of silver eventually being transferred from the counterpart. One can conclude that wear progresses mainly by plastic deformation of the soft polymer material. This can be further confirmed in Fig. 6b showing the smearing of the PE ridges on the scar edge.

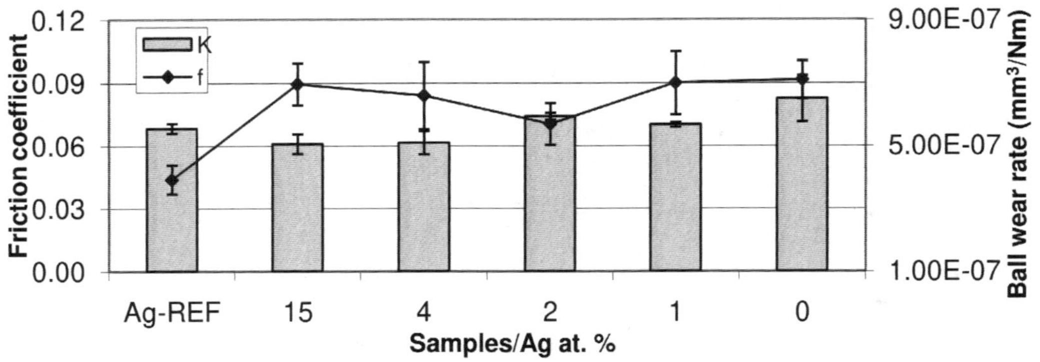

Figure 4. Friction coefficient and ball wear rates for the Ag-doped diamond-like carbon coatings against PE balls. The results obtained with a pure silver and amorphous carbon coating are included for comparison.

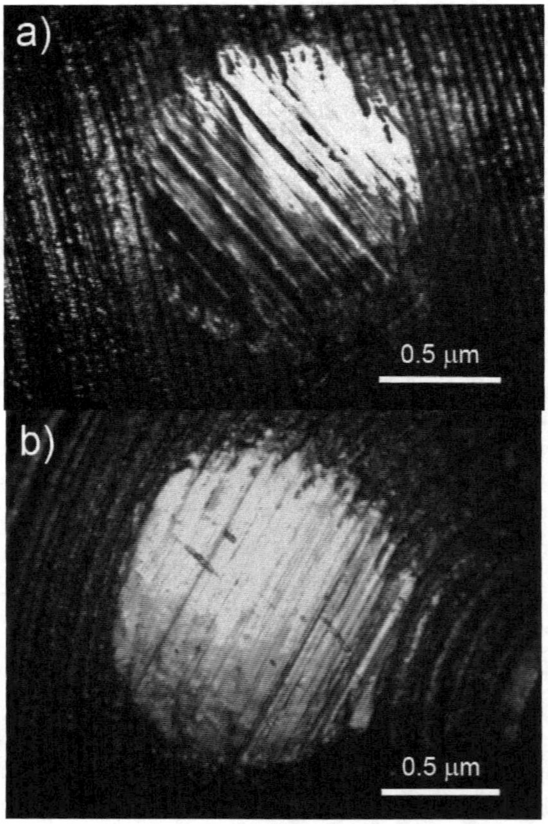

Figure 5. Optical micrographs taken from the ball scars after testing the following samples: a) Ag (15 at.%)-DLC and b) undoped a-C.

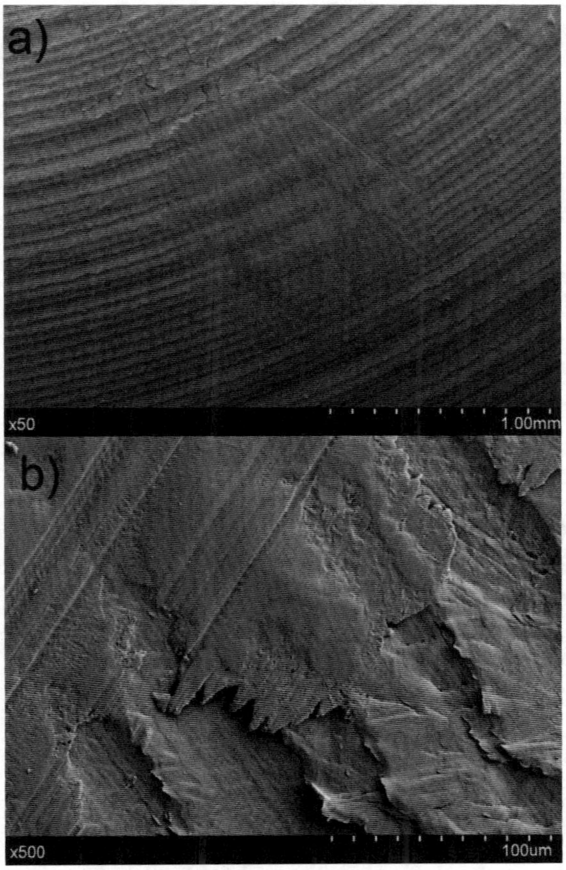

Figure 6. SEM micrographs from UHMWPE ball rubbed against the Ag film: a) wear scar; b) detail on the edge.

4. Conclusions

In this study, the influence of the silver content on the tribological properties of hydrogen-free Ag-DLC films against UHMWPE lubricated with fetal bovine serum was investigated. The incorporation of silver into DLC by means of "selective bias" pulsed dual-cathode cathodic arc results in nanocomposite coatings composed of nanocrystalline silver grains embedded in an amorphous carbon matrix, as indicated by both XRD and Raman spectra, where the structural characteristics of the amorphous carbon matrix are independent of silver content. There was a slight reduction in the wear volume of the UHMWPE balls for DLC containing 4 at.% or more of silver. SEM investigations of wear ball scars indicated this effect is due to a less abrasive interaction between the UHMWPE ball and the softer surface of silver-containing DLCs.

Acknowledgements

The authors thank J.M. Albella (ICMM, CSIC) for his helpful comments, A. Espinosa and A. de Andrés for the collection of the Raman spectra and M.C. Jiménez de Haro for taking the SEM pictures. This work was partially supported by MICINN project CSD 2008-0023 FUNCOAT. A.A. acknowledges support by the U.S. Department of Energy under Contract No. DE-AC02-05CH11231. JLE also acknowledges Spanish MICINN Project FIS2009-12964-C05-04.

References

[1] Robertson J 2002 *Materials Science and Engineering: R: Reports* **37** 129-281
[2] Allen M, Law F and Rushton N 1994 *Clinical Materials* **17** 1-10
[3] Dearnaley G 1993 *Clinical Materials* **12** 237-44
[4] Dearnaley G and Arps J H 2005 *Surface and Coatings Technology* **200** 2518-24
[5] Rodil S E, Ramirez C, Olivares R, Arzate H, Reyes-Gasga J and Magana C 2006 *Diamond and Related Materials* **15** 1300-9
[6] Tiainen V-M 2001 *Diamond and Related Materials* **10** 153-60
[7] Endrino J L, Marco J F, Allen M, Poolcharuansin P, Phani A R, Albella J M and Anders A 2008 *Applied Surface Science* **254** 5323-8
[8] Endrino J L, Escobar Galindo R, Zhang H S, Allen M, Gago R, Espinosa A and Anders A 2008 *Surface and Coatings Technology* **202** 3675-82
[9] Endrino J L, Allen M, Escobar Galindo R, Zhang H, Anders A, Albella J M 2006 *Materials Research Society Symposium Proceedings* **950** 3675-3682
[10] Lungu C P 2005 *Surface and Coatings Technology* **200** 198-202
[11] Morrison M L, Buchanan R A, Liaw P K, Berry C J, Brigmon R L, Riester L, Abernathy H, Jin C and Narayan R J 2006 *Diamond and Related Materials* **15** 138-46
[12] Narayan R J 2005 *Diamond and Related Materials* **14** 1319-30
[13] Wang D-Y, Chang Y-Y, Chang C-L and Huang Y-W 2005 *Surface and Coatings Technology* **200** 2175-80
[14] Willert H G, Bertram H and Hans Buchhorn G 1990 *Clinical Orthopaedics and Related Research* **95**-107
[15] Anders A, Brown I G, MacGill R A and Dickinson M R, 1998 *J. Phys. D: Appl. Phys.* **31** 584-7
[16] Anders A, Pasaja N and Sansongsiri S 2007 *Rev. Sci. Instrum.* **78** 63901
[17] Casiraghi C, Ferrari A C and Robertson J 2005 *Physical Review B (Condensed Matter and Materials Physics)* **72** 085401-14
[18] Ferrari A C and Robertson J 2000 *Physical Review B (Condensed Matter and Materials Physics)* **61** 14095-107
[19] Narayan R J, Wang H and Tiwari A 2003 *Materials Research Society Symposium Proceedings* **750** 205-210
[20] Kelly P J, Li H, Whitehead K A, Verran J, Arnell R D and Iordanova I 2009 *Surface and Coatings Technology* **204** 1137-1140

Tribological performance of DLC coatings on UHMWPE

V Martínez-Nogués[1,*], F J Medel[2], M D Mariscal[1], J L Endrino[3], J Krzanowski[4], F Yubero[5] and J A Puértolas[1,2]

[1]Department of Materials Science and Technology, Instituto de Investigaciones en Ingeniería de Aragón, I3A, Universidad de Zaragoza, C/ María de Luna 3, E-50018 Zaragoza, Spain.
[2] Instituto de Ciencia de Materiales de Aragón, ICMA, Universidad de Zaragoza-CSIC, E-50018 Zaragoza, Spain
[3] Instituto de Ciencia de Materiales de Madrid (CSIC). Sor Juana Inés de la Cruz, 3, Cantoblanco, E-28049 Madrid, Spain.
[4] Materials Science Program and Department of Mechanical Engineering. University of New Hampshire, Durham, NH 03824 USA
[5] Instituto de Ciencia de Materiales de Sevilla (CSIC- U. Sevilla). C/Américo Vespucio 49, E-41092 Sevilla, Spain.

*Email: enav@unizar.es

Abstract. Diamond like carbon (DLC) coatings were deposited by several methods (ion beam assisted evaporation, magnetron sputtering, filter cathodic arc, and plasma enhanced chemical vapor deposition) onto medical grade ultra-high-molecular weight polyethylene (UHMWPE) discs. The chemical characteristics and mechanical properties of the deposited DLC coatings were studied by Raman spectroscopy and nanoindentation, respectively. In addition, a set of tribological tests was conducted at human body temperature and under bovine serum lubrication against alumina balls. After testing, wear tracks were both visually inspected and documented using confocal microscopy. Visual inspection of the wear tracks confirmed that the DLC coatings were completely removed in all cases, the only exception being the DLC coating prepared by magnetron sputtering with thickness about 0.5 microns. Although this type of DLC coating exhibited the highest friction coefficient, and therefore it suggested a somewhat lower resistance to abrasive/adhesive wear conditions, no evidence of cracking or delamination was observed after the high contact pressure wear testing. This fact points out a good substrate-coating adhesion, and confirms magnetron sputtered DLC as a potential coating for orthopaedic applications.

1. Introduction

Ultra high molecular weight polyethylene (UHMWPE) is the "gold standard" for bearing surface in total joint arthroplasty due to its physical, chemical, and mechanical properties, as well as biocompatibility. However, UHMWPE is the softer material in total joint replacements and can exhibit wear-related problems and mechanical degradation after long-term post irradiation oxidation. The main problem is the generation of particles from its use in both articular and non-articular surfaces, which has been associated with osteolysis and implant loosening. The wear of UHMWPE on non-articular surfaces is also known as secondary wear and it is caused by micromobility in modular junctions due to insufficient locking during service. Although it was first detected in hip replacements, this problem affects knee replacements with polyethylene particle production estimations being 2-100 times the amount for femorotibial wear [1].

A solution to this problem could be to coat non-articular surfaces of polyethylene implants with low friction and wear resistant films. Diamond- Like-Carbon (DLC) coatings are potential candidates to reduce UHMWPE wear debris generation due to their excellent mechanical properties (hardness, wear resistance, friction), high chemical inertness and good biocompatibility. [2,3]

In this study, we compared the tribological behavior of DLC coatings obtained by several deposition methods: ion beam assisted evaporation (IBAD), magnetron sputtering (MS), filter cathodic arc (FCA) evaporation, and plasma enhanced chemical vapor deposition (PECVD). The objective of this work was to determine the potential use of DLC coated UHMWPE to reduce the wear debris formation in non-articular surfaces.

2. Materials and methods

2.1. UHMWPE sample preparation

An extruded GUR 1050 UHMWPE (Perplas Medical Ltd., Lancashire UK) bar was used as raw material in the preparation of 20 mm diameter discs (3 mm thick). All disks were ground and polished using SiC papers up to a surface roughness equal to 0.15 ± 0.05 μm, measured using a SENSOFAR PLu 2300 optical imaging profiler.

2.2. DLC coatings

The DLC films were deposited on pristine UHMWPE by several techniques: IBAD, MS, FCA and PECVD. The samples included in this study were labelled as M1-M7, as shown in Table 1, where their thickness and average roughness R_a are included.

Table 1. Thickness and average roughness Ra of the DLC coatings on UHMWPE considered in this work

Sample	Deposition method	Thickness (μm)	R_a (μm)
M1	IBAD	0.10	0.96
M2	IBAD	0.50	2.86
M3	MS	0.25	0.10
M4	MS	0.49	0.18
M5	FCA	0.13	0.15
M6	FCA	0.10	0.10
M7	PECVD	0.70	0.20

Samples M1 and M2 were synthesized at room temperature using the IBAD technique. The evaporation was performed from a single graphite source in combination with Ar^+ bombardment using a Kaufmann type source. Likely, the high roughness observed in these samples was due to localized heating effects induced by the Ar^+ ions at the surface of the UHMWPE substrates. Samples M3 and M4 were prepared using MS from a pure graphite target at room temperature with grounded substrates. Samples M5 and M6 were synthesized using a “triggerless” mini-gun designed to operate in pulsed mode. The pulses used in the production of carbon plasma were 100 Å and had a duration of 5 ms per second, the plasma stream produced by the source is injected into a 90-degree filter to remove most of the macroparticles which were formed during the cathodic arc process. During the deposition of the FCA DLC samples, the substrates were pulsed polarized at -1000 V with a duty cycle of 15%. PECVD technique was made in a RF capacitive coupled reactor with plate parallel electrodes. DLC layers of 700 nm were deposited on pristine substrates using acetylene, Ar and H_2 gas mixture and 200 V bias voltage.

The local bonding characteristics of the DLC films were studied with Raman spectroscopy with visible (523 nm) excitation. Spectra were collected using a Horiba Jobin Yvon HR800 system equipped with a CCD camera and a confocal microscope. The laser spot was focused on the sample surface using a 50x objective with short-focus working distance.

2.3. Nano-hardness profiles

In- depth hardness profiles of coated samples were obtained using a nanoindenter XP MTS system. The maximum load used in the tests was 2.0 mN. The indentation rate was 5 nm /s.

2.4. Wear experiments

A ball-on-disk tribometer (CSM instruments; Peseux, Switzerland) allowed assessing wear resistance and monitoring the coefficient of friction for each system. Wear tests were performed in a sample per material group working with a rotation vessel which contains UHMWPE disks immersed in bovine serum (B-9433, Sigma Aldrich). The counterpart was a stationary 6 mm diameter ball made of alumina with a nominal roughness of 0.050 ± 0.002 μm, whereas the load applied to the ball was 5.23 N. The radius of the circular track was 4 mm and the sliding speed was 0.05 m/s. The environment temperature was set to 37 ºC, to simulate body temperature. Assuming hertzian contact, the previous loading conditions gave an approximate contact pressure of 37 MPa, which is in the range of peak contact stresses (≥30 MPa) found for some contemporary tibial inserts. Wear rates were measured after 26400 meters sliding distances. After each wear test, disks were removed from the tribometer and underwent a cleaning protocol following guidelines included in ASTM F2025.

Confocal microscopy was performed using a SENSOFAR PLU 2300 optical imaging profiler to evaluate worn disk surfaces. A total of four line measurements diametrically opposed were carried out on the wear tracks of each sample. The average of the registered worn areas was used to calculate the corresponding wear factor.

The volume loss corresponding to the track profile was evaluated taking into account the plastic deformation suffered by the polyethylene disks under applied loads during wear tests. By virtue of the shape memory effect, UHMWPE heated to high enough temperature recover the original shape that has been previously lost due to mechanical deformation. The recovery of plastic deformation after wear testing was achieved by heating samples at 120 ° C for 36 hours. Preliminary tests confirmed that some DLC coated samples demonstrated no removal of the DLC layer after tribological testing, but they exhibited wear tracks exclusively due to permanent plastic deformation. This permanent plastic deformation was not completely recovered after the thermal treatment. Thus, the average worn area after the heating process was corrected taking into account the non-recoverable plastic deformation arise in the sample where the DLC layer remains.

3. Results and discussion

3.1. Chemical characterization of the DLC coatings

Figure 1 show the typical Raman spectra registered for the thickest DLC-coated UHMWPE samples considered in this study.

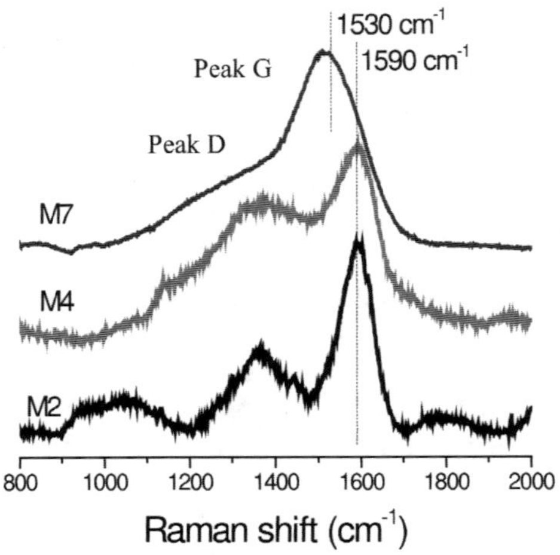

Figure 1. Typical Raman spectra corresponding to thick DLC coatings deposited by several techniques on UHMWPE.

The Raman spectra exhibited the characteristic G (due to the bond stretching of all pairs of sp^2 atoms in rings and chains) and D (breathing modes of sp^2 atoms in rings) peaks at 1530-1590 cm^{-1} and at 1330 cm^{-1}, respectively. The spectra were analysed by peak fitting and a linear background subtraction. Three values were extracted from the fitting process, the position w_G and full-width at half maximum Δ_G of the G peaks, and the ratio between intensities of the D and G peaks, I_D/I_G. The first parameter w_G is mainly related to the sp^3 content in the DLC material. On the other hand, Δ_G is sensitive to structural disorder (bond angle and bond length distortion), and I_D/I_G is related to the size of the sp^2 clusters organized in rings [4, 5]. The values obtained for the thickest DLC coatings considered in this study are shown in Table 2 (Raman analysis of the thinnest coatings were not attempted due to strong interferences with the signal originated at the UHMWPE substrate).

Table 2. Raman parameters characteristic of the thickest DLC coatings considered in this work

Sample	M2	M4	M7
	IBAD	MS	PECVD
w_G (cm^{-1})	1589	1594	1542
Δ_G (cm^{-1})	77	116	130
I_D/I_G	0.87	0.71	0.66

The coatings prepared by IBAD and MS showed values of w_G of about 1590 cm^{-1}. These high values are usually correlated with films with high content of nanocrystaline graphite [4, 5]. The corresponding values of Δ_G are consistent with the presence of sp^2 clusters smaller than 2 nm. Regarding the film prepared by PECVD, the value of w_G is consistent with highly disordered amorphous hydrogenated DLC (about 20% at.) with 30-40% of sp^3 bonds [4, 5].

3.2. Nano-hardness

Figure 2 shows the nanoindentation hardness depth profiles obtained for DLC coated UHMWPE samples. Qualitatively, a similar behavior was observed for all the DLC coatings. The influence of the soft UHMWPE substrate was typically observed at indentation depths as shallow as 20 nm. Although M7 (PECVD) showed slightly higher hardness profile values than DLC coatings obtained using other deposition methods, this could be due to its higher thickness since its initial surface hardness value is one of the lowest (~3 GPa) of the series. Nano-hardness tests performed on MS samples (M3 and M4) had the highest start-up value of ~26 and ~15 GPa (range not shown in Fig. 2), respectively with a further decrease up to around 0.2 GPa. The difference in the measured hardness between M3 and M4 could also be related to the difference in thickness and the influence of the soft substrate since the same tendency appears in M5 and M6 with the thicker coating showing the higher hardness values. The sample deposited by evaporation (M1) had an initial surface hardness close to 10 GPa and its hardness profile decreased to the value expected for virgin UHMWPE (~0.01GPa) for depths shallower than those of the rest of the films. Nanoindentation depth profile values could not be collected for M2 sample due to its high surface roughness.

In agreement with [6], sample M7 had the lowest I_D/I_G ratio, so its sp^3 content is higher than in the rest of the samples. Due to this fact it shows better mechanical properties (higher hardness) during the measuring profile, as we shown in Figure2.

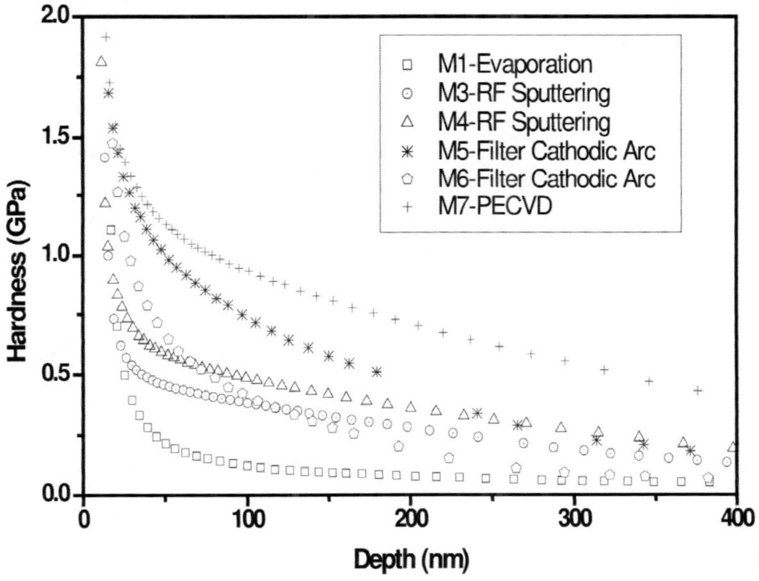

Figure 2. Hardness profiles of the DLC coatings deposited on UHMWPE

3.3. Friction

The friction coefficient versus sliding distance obtained during ball-on-disk tests are shown in Fig. 3. Sample M5 (FCA) presented the lowest friction coefficient (~0.10) at the beginning of the run and remained practically constant after a sliding distance of 25000 meters. This value is similar to that obtained for the uncoated UHMWPE substrate using identical tribological conditions [7]. The sample deposited using MS displayed the highest friction coefficient, around 0.30, and it remained above 0.25 for the entire run, indicating that the coating was not removed from the substrate during the test. The evolution of the friction coefficient of the samples obtained using PECVD and IBAD methods revealed a different tribological behaviour. Both of them started at 0.25, but the friction coefficient values tend to decrease during the test to 0.20 and 0.15 respectively after the running-in period. This points out that the DLC coatings deposited by these two methods worn out after a sliding distance of 16000 and 6000 m, respectively. Likely, the same mechanism occurred in the sample deposited by FCA. In this case, the thickness of the coating was so thin that was probably partially removed during the running-in stage yielding a low friction coefficient.

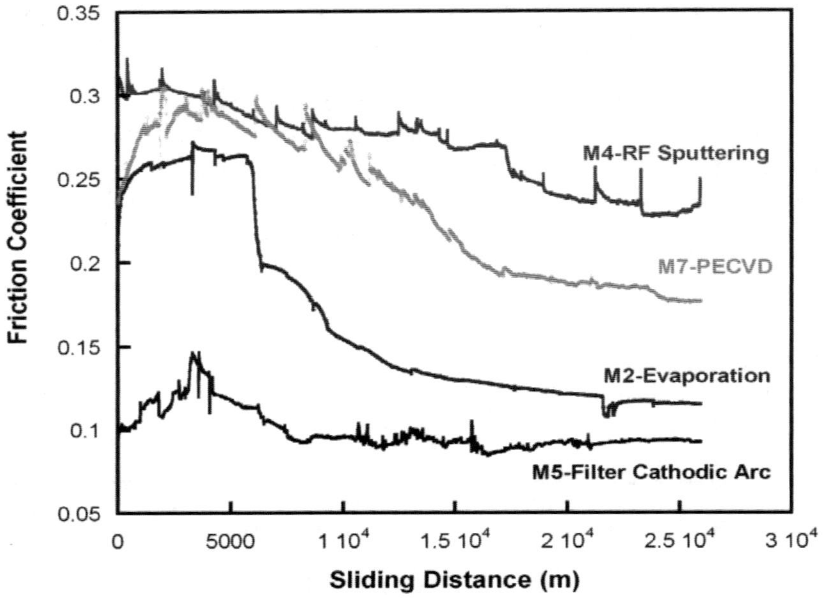

Figure 3. Evolution of the average friction coefficients during the ball on disk tests for DLC coated UHMWPE disks.

3.4. Wear

Confocal microscopy images of the wear tracks (Fig. 4a-g) were obtained for the seven coated UHMWPE discs after tribological testing (26400 meters of sliding distance). All the samples showed wear tracks featuring grooves parallel to the sliding direction. These grooves are typical of an abrasive wear process, likely produced by the contact between alumina ball and DLC asperities. In the case of M4, the only sample in which the coating remained practically intact after the entire test, the micrograph showed a lighter colour, due to the presence of the DLC layer, but still similar abrasive grooves. However, the wear track of the PECVD sample showed a mixed behaviour since it was partially covered by DLC fragments, indicating that, in this case, the DLC coating had been fragmented and partially removed in some areas.

The wear performance of the samples included in this work is summarized in Table 3. After the recovery process, some samples, M5 and M7 (NW-PE), showed a negligible wear of the PE when the DLC coatings were removed from the substrate.

The most relevant result was that the MS coating with ~0.5 μm in thickness, M4 (NW-DLC), did not exhibit signs of fragmentation or coating delamination after 26000 m of sliding distance in the described tribological conditions. The performance was better than that shown for other DLC coated UHMWPE samples with similar thickness but obtained by other methods (IBAD or PECVD). The joint interpretation of friction and wear results confirmed that, despite the comparatively lower abrasion resistance as reflected by the high friction coefficient, the 0.5 μm thick MS DLC layer presented a strong substrate-coating adhesion. Thus, M4 was able to endure the high contact pressure conditions and the predominant fatigue wear mechanism of the present tribological test.

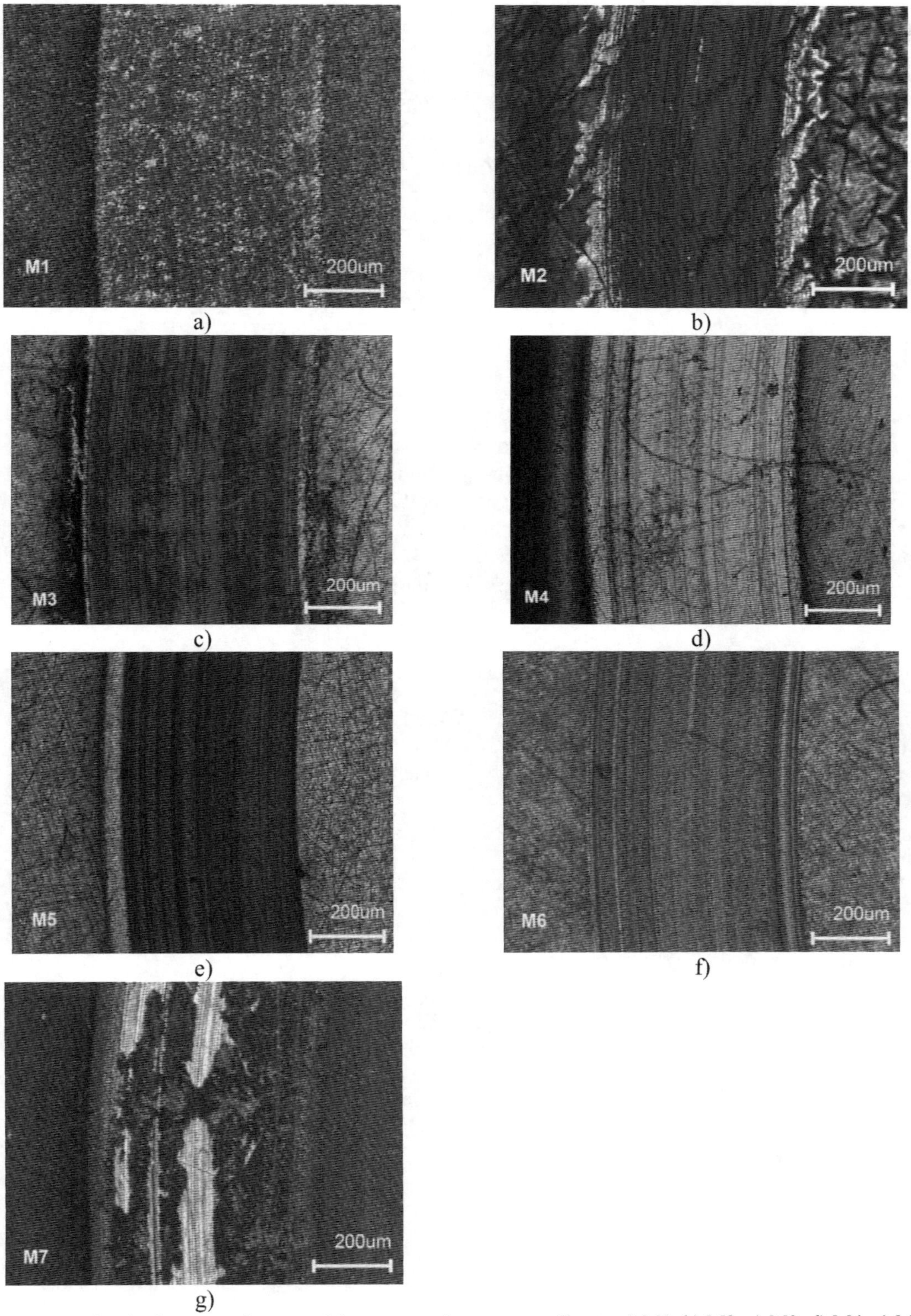

Figure 4. Confocal microscopy images of the wear tracks corresponding to a) M1, b) M2, c) M3, d) M4, e) M5, f) M6 and g) M7.

Table 3. Wear rate of DLC coated materials after 26400 m sliding distance, where NW-DLC means samples without DLC wear and NW-PE means no measurable wear over the PE.

Sample	**Wear rate x10^{-7} (mm^3/Nm)**
M1	1.1 ± 1.0
M2	2.5 ± 1.8
M3	1.7 ± 0.3
M4	NW-DLC
M5	NW-PE
M6	0.7 ± 0.5
M7	NW-PE

4. Conclusions

Remarkable wear test results were obtained for DLC-coated UHMWPE disc prepared using the radio frequency magnetron sputtering (RF-MS) technique with high content of nanocrystalline graphite and thickness of ~0.5 μm. Even though the friction coefficient of these MS DLC-coated UHMWPE samples were the highest of the set. The excellent wear behavior observed, with no delamination or chipping even after 26000 meters sliding distance, confirms the potential of these nanocrystalline graphite rich DLC coatings on UHMWPE for some orthopedic applications.

Acknowledgment

The authors are grateful to A.Anders (LBNL) and I. Jimenez (ICMM) for deposition of DLC coatings by filter cathodic arc and by IBAD techniques. We thank the Ministry of Science and Education of Spain (projects MAT2006-12603-C02-01 and CONSOLIDER-INGENIO CDS2008-0023 for financial support. JLE also acknowledges Spanish MICINN Project FIS2009-12964-C05-04.

References

[1] A.R. Rao, G.A. Engh, M.B. Collier, S. Lounici. J.Bone Joint Surg. Am. 84-A (10) (2002) 1849 1855.
[2] R.K. Roy, K.R.Lee. J.Biomed Mater. Res.B: Appl. Biomater. 83(1)(2007) 72-84.
[3] R. Hauert. Tribol.Int 37 (11-12) (2004) 991-1003.
[4] J. Robertson, Materials Science and Engineering R 37 (2002) 129.
[5] C. Casiraghi, A.C. Ferrari, J. Roberson, Phys. Rev. B72 8 (2006) 08540.
[6] F.X. Liu, Z.L. Wang. Surface coating Technology. 203 (2009) 1829-1832.
[7] J.A. Puértolas, V. Matínez-Nogués, M.J. Martínez-Morlanes, M.D. Mariscal, F.J. Medel, C. López-Santos, Y. Yubero. Wear 269 (2010) 458-465

Journal of Physics: Conference Series **252** (2010) 012007 doi:10.1088/1742-6596/252/1/012007

Multifunctional sol-gel derived thin film based on nanocrystaline hydroxyapatite powders

A.A. El hadad [1], V. Barranco[2], A. Jiménez-Morales[3], E. Peon[4], J.C. Galván[1]

[1]Centro Nacional de Investigaciones Metalúrgicas (CSIC), Madrid, Spain. [2]Instituto de Ciencia de Materiales de Madrid (CSIC), Madrid, Spain. [3]Universidad Carlos III de Madrid. Departamento de Ciencia e Ingeniería de Materiales e Ingeniería Química. (Madrid), Spain. [4]Centro de Biomateriales, Universidad de La Habana, Havana, Cuba

E-mails: amirelhadad@cenim.csic.es, jcgalvan@cenim.csic.es

Abstract. The aim of this work was to prepare bioactive hydroxyapatite coatings by sol-gel method and to study the effect of thermal treatment temperature upon the bioactivity and corrosion protection of these coatings on Ti6Al4V alloy. The application of (DTA/TGA) and (XRD) has provided valuable information about the phase transformation, mass loss, identification of the phases developed, crystallite size and degree of crystallinity. (SEM/EDX) has been applied to study the surface morphology of coated samples before and after immersion in simulated body fluid (SBF) to detect the biomimetic precipitation of the bone-like apatite. The obtained results show that all the prepared samples are ceramic nanocrystalline with crystal structure and composition like hydroxyapatite, with little deviations from that present in the human bone. The bioactivity of the studied samples is found to be closely related to the thermal treatments applied. That is, the bioactivity decreases as the temperature of the thermal treatment increase. Coatings from such prepared hydroxyapatite sol have been accomplished by dip-coating technique on non-toxic Ti6Al4V alloy for biomedical applications. The corrosion behaviour of the resulting hydroxyapatite coatings in a (SBF) has been studied by electrochemical impedance spectroscopy (EIS). The hydroxyapatite coated Ti6Al4V alloy displayed excellent bioactivity when soaked in the (SBF) and acceptable corrosion protection behaviour.

1. Introduction

The bone tissue is considered as a natural composite, composed of two constituents organic and inorganic solid, the inorganic solid is carbonated hydroxyapatite (CHA) while the organic one is collagen. The hydroxyapatite (HA) which has the formula $Ca_{10}(PO_4)_6(OH)_2$, amounts to 65% of the total bone mass, with the remaining mass formed by organic matter "mostly collagen" and water[1]. The collagen molecules are bonded together forming fibrils, which bound together to form fibers. Between these fibers there are small interstitial empty spaces, where apatite nanocrystals are deposited. For this reason, the most used calcium phosphate in implant fabrication is HA, a biological active material with different forms, particles, films, coatings, fibers which has extensive biomedical applications [2]. Nano-sized HA powders have a high specific surface area and therefore exhibit enhanced activity toward chemical and biological interactions in the human body. HA has been known to form *in vitro* and *in vivo* a bioactive bone-like apatite layer spontaneously on its surfaces. The formed apatite layer act as interface between the implant and tissue, facilitate the formation of chemical and biological bond with the osseous tissue. Therefore, the essential prerequisite for a biomaterial to bond directly to living bone is the formation of an apatite layer on its surface when implanted in the body. As this layer is formed, the protein absorption onto the surface of such layer will trigger the osteoblasts, "new bone forming cells", to differentiate and reproduce, leading the formation of new bone tissue.

The HA can be synthesized by many methods [3,4]. Among all of them, the sol-gel process is the most used technique for production of HA because it offers many advantages over other traditional methods. It allows the production of ultra fine and pure ceramic powders with relatively low cost, low synthesis temperature, ability to coat porous and complex surfaces with high control of composition and microstructure of the obtained film [5-7]. The HA synthesized by sol-gel will form strong chemical bonds with bone *in vivo*, all of these mentioned properties place the HA into a class of biomaterials known as bioactive materials. Unfortunately brittleness limits its usage for load bearing applications. For that it can only be used as bone fillers and bone graft substitutes in orthopedics[8]. The only exception where HA is applied in dynamically loaded situations is the use of HA as a coating material.

In the present study, HA was synthesized by adopting a sol–gel approach. The resultant HA was characterized using DTA, TGA, XRD and SEM/EDX techniques. Sol–gel-derived HA coatings on Ti6Al4V substrate were evaluated via electrochemical studies involving impedance measurements in Kokubo's simulated body fluid (SBF) solution.

2. Experimental

2-1. Preparation of Hydroxyapatite Powders.

The prepared HA sol-gel was obtained by hydrolysis and condensation of two precursors. Triethyl phosphite, $(C_2H_5O)_3P$ (Fluka) and calcium nitrate tetrahydrate, $Ca(NO_3)_2{\cdot}4H_2O$ (Panreac) were employed as phosphorus and calcium precursors respectively. In brief, 0.03mol of triethyl phosphite was first diluted in ethanol and hydrolyzed for 24 h with a fixed amount of distilled water in a parafilm-sealed glass container under vigorous stirring.

A stoichiometric amount, 0.05 moles, of $Ca(NO_3)_2{\cdot}4H_2O$ is dissolved first in 25 ml of anhydrous ethanol then added dropwise into the hydrolyzed phosphite sol. The mixed sol solution was then continuously agitated for additional 30 min and kept static (aging) at 45°C. The pH value was adjusted to be 7.5; the sol was aged static at 60°C for 16 h. The solvents were then driven off resulting in a viscous liquid. Further drying at 80°C was done to obtain the prepared material in a powder form. The obtained powder was placed in a steel die, and pressed to 7 GPa to produce 12.9 mm diameter pellets which were was sintered at various temperatures (600, 800, and 1200°C).

2-2. Preparation of HA Coatings on Ti6Al4V Substrates

Ti6Al4V disks of 2 cm of diameter and 0.4 cm of thickness were polished using different silicon carbide grit up to 1200 grade. The substrates were ultrasonically degreased with acetone for 10 min and washed with distilled water. Finally, the substrates were dried at 200°C for 1 hour in an air oven to form a titanium oxide surface. These substrates were dip coated with the HA sol solution, with a dipping and withdraw speed of 12 cm/min. The sol-coated substrates were then immediately transferred into an air oven and held at 80°C for 30 min. To increase the thickness of the coating, the above process was repeated for three times and finally it was sintered at various temperatures (600, 800°C).

2.3 Characterization of HA Powder Samples

2.3.1 Thermal Behaviour (DTA/TGA)

The thermal behavior and mass loss of the powder was carried out using (**SETARAM DTA-TG** Setsys Evolution -1750), with α-Al_2O_3 powders as a reference material. HA powder samples with a weight of about 50 mg were analyzed. The samples and the reference material were heated in air from ambient temperature to 1300°C in argon at a heating rate of 20°C/min.

2.3.2 X-ray Diffractometer Analysis (XRD)

X-ray diffraction patterns have been recorded using a Siemens diffractometer D5000 equipped with scanning rate 0.1° in the 2θ ranging from 10° to 100° step time 1 sec. The prepared HA powder samples (S1 "as-prepared", S6 "treated at 600°C", S8 "treated at 800°C" and S12 "treated at 1200°C") were examined with a high-resolution X-ray powder diffractometer using monochromatic Cu Kα radiation (λ= 0.15406 nm). The results have been used to notice the change of developing phases, crystal structure of HA and degree of crystallinity, with the variation of heat treatment temperatures.

2.4. Evaluation of Bioactivity and Corrosion Protection

2.4.1 Immersion Tests of HA-Coatings/Ti6Al4V System in SBF

The *in vitro* bioactivity of the prepared HA-coatings/Ti6Al4V samples was tested by soaking in Kokubo's simulated body fluid (SBF). The pH of the SBF solution was buffered at 7.4. The specimens were immersed in separate plastic containers then incubated for 15 days at 37.5°C. After 15 days the specimens were removed from the solution, rinsed with distilled water and left to dry at room temperature.

2.4.2 Corrosion Behaviour

The corrosion behaviour of the HA-coatings/Ti6Al4V system was evaluated by applying electrochemical impedance spectroscopy measurements (EIS). For this purpose a potentiostat/galvanostat AutoLab EcoChemie PGSTAT30 (Eco Chemie, Utrecht, The Netherlands) equipped with a FRA2 frequency response analyzer module was used. A three electrode setup was applied. The working electrode was the investigated sample with an area of 2.14 cm^2. The reference and the counter-electrode were a saturated calomel electrode (SCE) and a large size graphite sheet, respectively. The electrochemical cell was filled with Kokubo's solution. The EIS measurements were made at the open circuit potential (OCP). Logarithmic frequency scans were carried out by applying sinusoidal wave perturbations of ±10mV in amplitude, in the range of 10^5-10^{-3}Hz. Five impedance sampling points were registered per frequency decade. The impedance data were analysed by using the 'EQUIVCRT' program developed by Boukamp (AC-Immittance Data Analysis System, Version 4.51, University of Twente, The Netherlands, 1993). The electrical resistance values were obtained by applying the fitting routines of the ZView® software version 3.1c (Scribner Associates Inc, Southern Pines, NC, USA).

2.4.3 Scanning Electron Microscopy (SEM/EDX)

The surface morphology of selected specimens, before and after immersion in SBF, was carried out by scanning electron microscopy (SEM) with a field emission gun (FEG) coupled with an energy dispersive X-ray (EDX) system (Oxford Inca microanalysis system and a windowless detector) for chemical analysis. The samples were examined in a JEOL-6500F microscope at 15 kV acceleration voltages. The samples were coated with ultra thin carbon conductive layer to avoid the charging due to the non conductive nature of the studied samples.

3. Results and discussion

Differential Thermal Analysis (DTA) and thermogravimetry (TG) diagrams of the obtained powders are given in Figure 1. The DTA was employed to determine the temperature at which the as-prepared powder, amorphous calcium phosphate (ACP), transforms to crystalline HA.

In the DTA curve of dry gel (powder), the first endothermic peak appeared at 38.8°C which was due to the evaporation of the absorbed water (residual moisture evaporation)[9]. The two endothermic peaks which appeared at 123.3°C and 443.2°C were attributed to elimination of crystalline water in un-reacted calcium nitrate tetra hydrate $Ca(NO_3)_2 4H_2O$ and removal of NO_3^- groups, as well as, to the condensation "dehydration" of hydrogen phosphate ions (HPO_4^{2-}) to form amorphous pyrophosphates ($P_2O_7^{4-}$) which is a component of amorphous calcium phosphates that forms during processing of HA synthesis[10-12]. The endothermic peak at 532.5°C is representative of the endothermic reaction indicating the crystallization of HA. The last endothermic peak observed at 1197.6°C may be attributed to the decomposition of the HA.

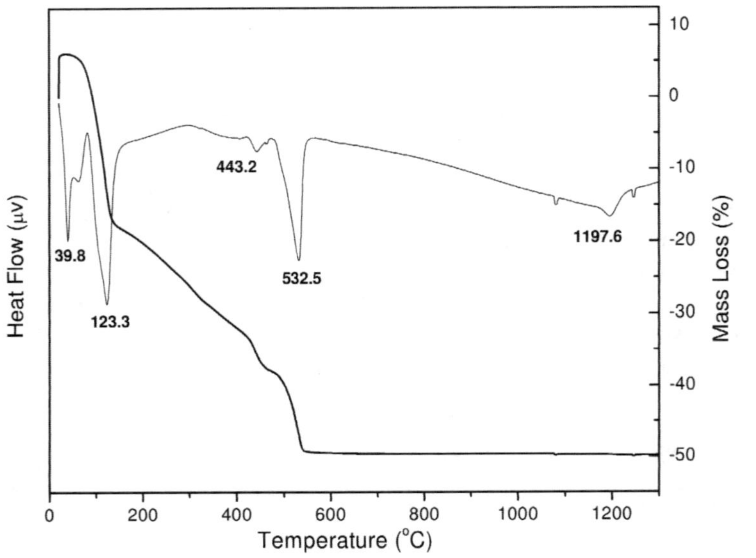

Figure 1. DTA/TG of the as prepared powder

In the TG, the weight loss was complete at about 570°C and three discrete weight-loss regions occurred as seen in the TG diagram. The first region at 44-172°C with 25% weight loss corresponds to the elimination of adsorbed water. This process was indicated as endothermic peak at 38.8°C and shoulder at 60°C in the DTA diagram. The second weight loss region appeared at 172-479°C in the TGA with 18% weight loss which corresponded to elimination of lattice water and was indicated as endothermic peak in the DTA diagram at 443°C. The third weight loss observed in the TG diagram at 479-570°C with a weight loss of 11 % corresponds to the decomposition of the nitrate salts which is also indicated as an endothermic peak at 532°C in the DTA curve.

X-ray diffractometry (XRD) identifying the developed phases, showed following features: a highly crystalline sample gives rise to narrow peaks and a poorly crystalline sample gives rise to broad peaks. "Crystallinity" refers to the degree of structural order in a solid. In a crystal, the atoms or molecules are arranged in a regular, periodic manner. In this respect means the degree of order within the crystal lattice [13]. According to Landi et al., the fraction of crystalline phase (X_C) can be calculated through the equation below [14]:

$$X_C = 1 - \frac{V_{112/300}}{I_{300}}$$

I_{300}: Intensity of peak diffracted from the (300) crystallographic planes of HA.
$V_{112/300}$: Intensity of the valley between the peaks of the planes (112) and (300) of HA.

The average crystallite size was calculated from the broadening in the XRD pattern. According to the Scherrer's equation[15]:

$$L_c = \frac{K\lambda}{\beta Cos\theta}$$

L_c: Average crystallite size (nm).
K: Shape coefficient (value between 0.9 and 1.0).
λ: Wavelength of X -ray beam - Cu Kα radiation (λ= 0.15406 nm).
β: Full width at half maximum (FWHM) of HA(211).
Θ: Diffraction angle

The diffraction peak corresponding to the (002) was chosen for calculation of the crystalline size, as it is relatively sharper than the other peaks.
Figure 2 illustrates the XRD patterns of S1 aged gel, dried at 80°C for 48 hours and of the "S6, S8, S12" samples thermally treated at 600, 800, 1200°C for 2 h, respectively. It can be interpreted by the discussion that sharp reflection peak appears in the range of 31.8–40° 2θ for the thermally treated samples "S6, S8, S12", representing the characteristic peak of HA apatite phase (JCPDS09-0432).

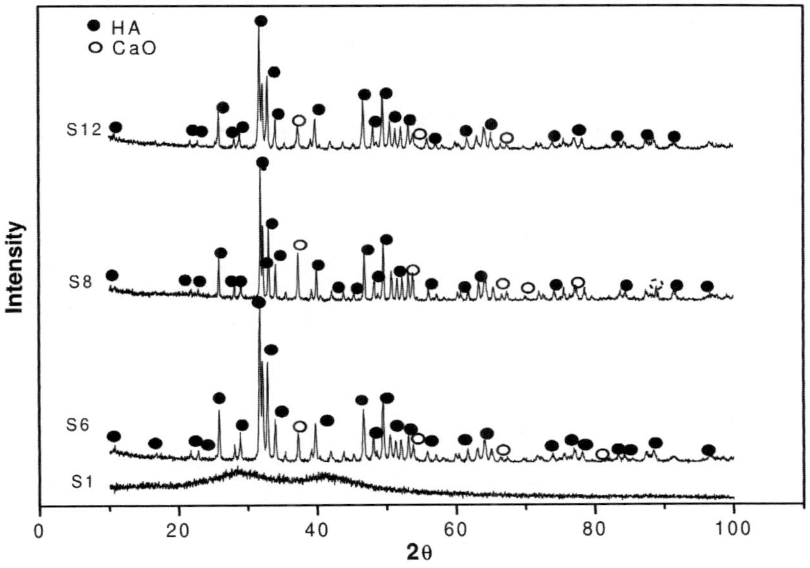

Figure 2. XRD patterns and developed phases of the dried gel (S1) and the powders thermally treated at different temperatures; 600°C (S6), 800°C (S8) and 1200°C (S12).

The XRD pattern of the as-prepared sample S1 is mainly amorphous, and there is no obvious diffraction peak observed. This fact is reported by many authors [16]. As the temperature increases, several peaks appear, which ascribed to the diffraction peaks of the HA phase. The XRD pattern of the S6, S8 and S12 sample thermally treated at 600, 800 and 1200°C respectively, indicates the existence of calcium oxide as traces, together with HA as main phase. The presence of CaO is due to the remaining $Ca(NO_3)_2$ in the precursor system, that can be directly transferred to CaO at high temperature[6]. The degree of crystallinity and calculated crystal size for the thermally treated samples can be seen in Table1. The degree of crystallinity of HA is increased when increasing the temperature from 600 to 800°C (Table 1), but decrease as the temperature increase to 1200°C. The decrease in crystallinity of the 1200°C thermally treated sample (S12) is attributed to the decomposition of HA at corresponding temperature 1200°C, in agreement with the results obtained from the DTA results (Figure 1), while the increase of the crystal size, when compared to 600°C and 800°C thermally treated samples is attributed to an increase of ion migrations when increasing the temperature (Table 1).

Table 1. Crystal size, degree of crystallinity, and electrochemical parameters of the powders thermally treated at different temperatures

Samples	Powders			Coatings		
	Thermal Treatment T(°C)	Degree of Crystallinity (%)	Crystallite Size (nm)	R_i /ohms·cm^2 (after 5 minutes in SBF)	R_f /ohms·cm^2 (after 15 days in SBF)	ΔR/Ri
S6	600	85	40	78610	134550	0.71
S8	800	89	51	5923	1208	-0.8
S12	1200	82	54	-	-	-

Electrochemical impedance spectroscopy (EIS) was used for evaluating the corrosion protection properties of the HA-coatings/Ti6Al4V system. Figure 3 shows the evolution of the electrochemical impedance spectra (Nyquist plots) with the immersion time in SBF for HA-coatings/Ti6Al4V system thermally treated at (A) 600°C/2h, (B) 800°C/2h, respectively. Initially, both samples exhibit very similar EIS behaviour. For short immersion time (5 min) the Nyquist plots draw very open arcs with high impedance values. Between 5 min and 1 day of immersion a fast decrease of the impedance values is observed which may be due to the penetration of the SBF through the pores present within the coating as a result of the evaporation of water and ethanol during thermal treatment. However, over time, the evolution of the impedance plots shows a different trend, which depends on the thermal treatment applied for preparing the HA coatings.

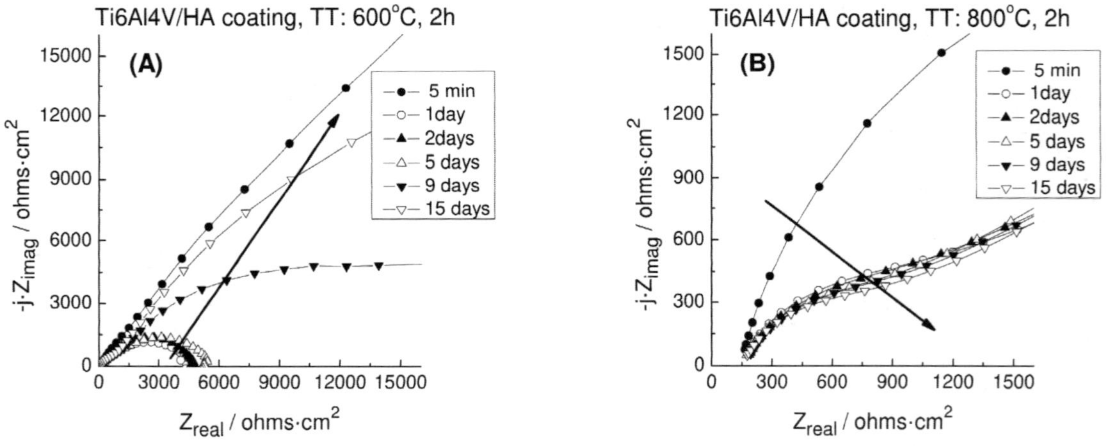

Figure 3 (a, b). Evolution of the electrochemical impedance spectra with the immersion time in SBF for HA-coatings/Ti6Al4V system thermally treated at (A) 600°C/2h, (B) 800°C/2h.

The sample thermally treated at 600°C/2hours exhibits a slow increase of the impedance between 1 day and 5 days. After that, the curvature of the Nyquist arcs are opening quickly and the total impedance reaches comparable values than those observed at the beginning of the immersion tests in the SBF. This behaviour can be due to the blocking of the pores by the formation of precipitates from the solution which prevent further penetrations. For the HA-coatings obtained at 800°C, the total impedance of the coating decrease continuously upon exposure to the SBF (Figure 3B) and the recovering of the impedance values observed in the samples coated at 600oC is not here evidenced, i.e. 15 days of immersion in the SBF are not enough to block all the pores present in the surface of the HA-coatings obtained at 800°C.

The evolution of the total resistance of the HA-coatings/Ti6Al4V thermally treated samples during the immersion time in Kokubo's simulated body fluid (SBF) was calculated from the impedance spectra by using the ZView software.

For the determination of the relative variation of total resistance of the metal/coating system, ($\Delta R/R_i$), the following equation was used:

$$(\Delta R/R_i) = \frac{R_f - R_i}{R_i}$$

where R_i is the initial resistance of the metal/coating system, R_f is the final resistance and, $\Delta R = R_f - R_i$.

Table 1 shows the values of the ($\Delta R/R_i$) coefficient obtained for thermally treated samples at 600 and 800°C, respectively. Samples treated at 600°C showed ($\Delta R/R_i$) coefficients with positive values (+0.71). This behaviour can be due to the blocking effect of the biomimetic precipitation of HA coming from the SBF which limits the liquid pathways to the metal surface. In contrast, the samples thermally treated at 800°C show a coefficient with negative values (-0.80) which can be due to the existence of surface cracks which are not able to be blocked by soaking these samples in the SBF.

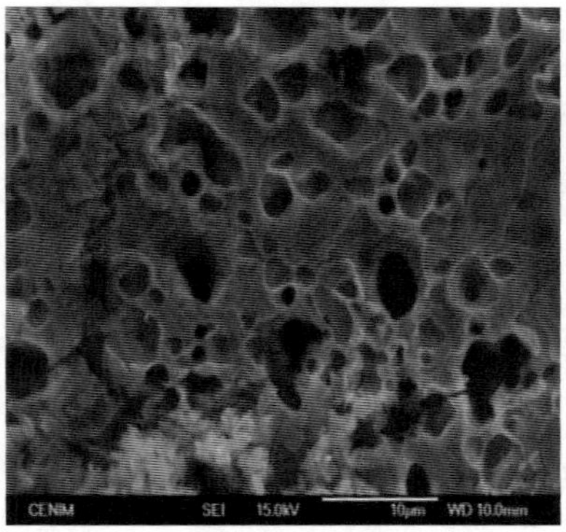

a)

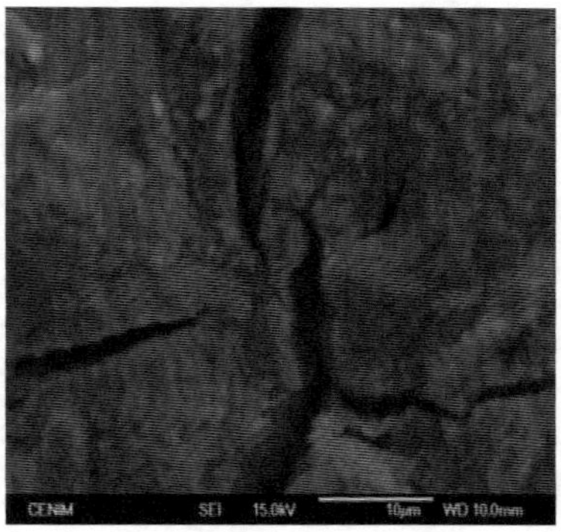

b)

Figure 4 (a, b). SEM micrographs obtained for the HA-coatings/Ti6Al4V system thermally treated at (a) 600°C/2h, (b) 800°C/2h before immersion in SBF.

The SEM micrographs obtained for the surface of the HA coatings before immersion in SBF are shown in Figure 4 (A, B). The surface of the sample thermally treated at 600°C (Figure 4A) shows the presence of pores due to the evaporation of both, water and ethanol during thermal treatment. While the surface of the sample thermally treated at 800°C (Figure 4B) show the presence of extensive cracks probably due to a thermal expansion coefficient mismatch between the coating and the substrates during cooling[17].

a)
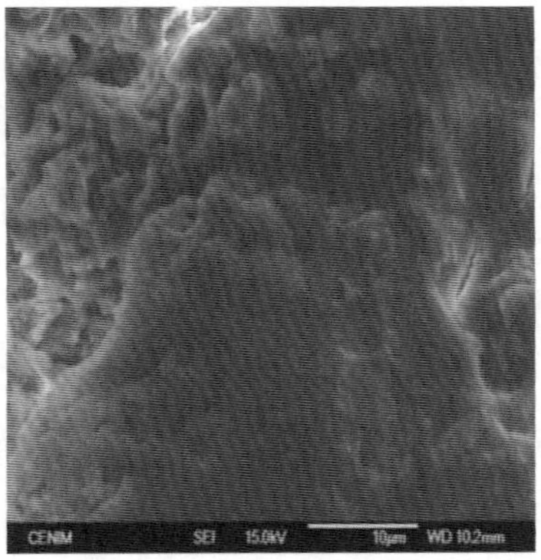

b)
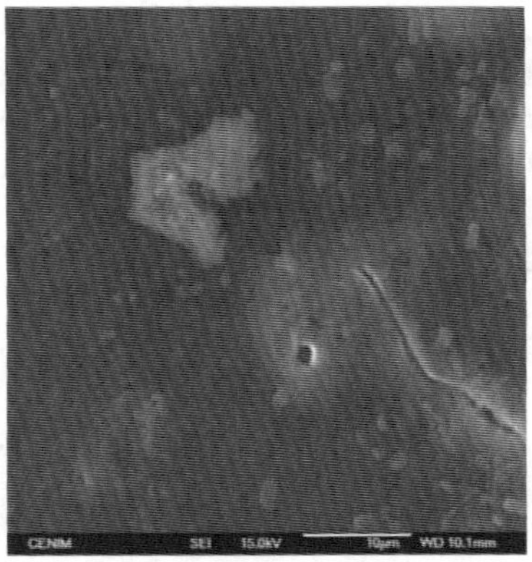

Figure 5. SEM micrographs obtained for the HA-coatings/Ti6Al4V system thermally treated at (a) 600°C/2h, (b) 800°C/2h after immersion in a SBF solution.

After 15 days of immersion in SBF the surface of both samples is covered with a bone-like apatite layer which proved that both samples are bioactive. The surface of the sample thermally treated at 600°C is completely covered with these precipitations from SBF (Figure 5A), while the surface of the sample thermally treated at 800°C is not completely covered (Figure 5B), as still some cracks are observed by SEM at the surface. These results are in agreement with those obtained by EIS.

5. Conclusions

- The crystallite size and degree of crystallinity of the HA sol-gel powders, which are precursors of the hydroxyapatite nano-crystalline coatings, strongly depends on the thermal treatment applied.
- Based on this knowledge, hydroxyapatite nano-crystalline coatings on Ti6Al4V alloy can be obtained by sol-gel route.
- Among all the coatings prepared, the HA coating/Ti6Al4V system obtained by thermal treatment at 600°C for 2 hours shows the best barrier properties against the *in vitro* corrosion of the metal substrate.
- All the prepared samples exhibit good bioactivity upon immersion in SBF at 37.5°C for 15 days. Among all of them, the 600°C thermally treated sample shows the best behaviour in terms of *in vitro* bioactivity.

Acknowledgements

This work has been supported by the National Program for Materials, Spanish Ministry of Science and Innovation (Project MAT2006-04486). A.A. El hadad acknowledges a pre-doctoral contract JAE financed by CSIC; V. Barranco acknowledges a Ramon y Cajal researcher contract financed by CSIC-MICINN.

References

[1] M. Vallet-Regi, J.M. Gonzalez-Calbet, *Prog. Solid State Chem.*, **32**, 1(2004).
[2] L.L. Hench, *J. Am. Ceram. Soc.*, **81**, 1705 (1998).
[3] A. Slosarczyk, E. Stobierska, Z. Paszkiewicz, M. Gawlick, *J. Am. Ceram. Soc.*, **79**, 2539 (1996).
[4] M. Yoshimura, H. Suda, K. Okamoto, K. Ioku, *J. Mater. Sci.*, **29**, 3399 (1994).
[5] D.M. Liu, T. Troczynski, W.J.J Tseng, *Biomaterials*, **22**, 1721(2001).
[6] M.F. Hsieh, L.H. Perng, T.S. Chin, H.G Perng, *Biomaterials,* **22**, 2601(2001).
[7] E. Peon, A. Jimenez-Morales, E. Fernandez-Escalante, M.C. Garcia-Alonso, M.L. Escudero, J.C. Galván, *Rev. Metal. Madrid*, Sp. Iss. 479 (2005).
[8] K. DeGroot, C.P.A.T. Klein, J.G.C. Wolke, J.M.A. De Blieck-Hogervorst, in "*CRC Handbook of Bioactive Calcium Phosphates*", edited by T. Yamamuro, L.L. Hench and J. Wilson, p. 3 (CRC Press, Boca Raton, FL, 1990).
[9] Diangang Wang, Chuanzhong Chen, Ting He, Tingquan Lei, *J. Mater. Sci.: Mater. Med.*, **19**, 2281(2008)
[10] Seok Kim, Prashant N. Kumta, *Mater. Sci. Eng. B.*, **111,** 232 (2004).
[11] D.G.Wang, C.Z. Chen, J. Ma, G. Zhang, *Colloid Surf. B-Biointerfaces*, **66**, 155 (2008).
[12] Kui Cheng, SAM Zhang, W.J. Weng, *J. Sol-Gel Sci. Technol.*, **38**, 13 (2006).
[13] M. Epple, *Z Kardiol*. 2001;90 Suppl 3:64-7.
[14] E. Landi, A. Tampieri, G. Celotti, S. Sprio, *J. Eur. Ceram. Soc.*, **20**, 2377 (2000).
[15] H.P. Klung, E. Alexander, "*X-Ray Diffraction Procedures: For Polycrystalline and Amorphous Materials*", 2nd Edition New York: John Wiley; (1974).
[16] Yanbao Li, Dongxu Li, *J. Appl. Ceram. Technol.*, **5**, 442 (2008).
[17] M.Wei, A.J.Ruys, B.K. Milthorpe, C.C. Sorrell, J.H. Evans. *J. Sol-Gel Sci. Technol* **21,** 39–48, (2001).

Journal of Physics: Conference Series **252** (2010) 012008 doi:10.1088/1742-6596/252/1/012008

Laser produced coatings and surface modifications for medical implants

B. León

Depto. de Física Aplicada, E.T.S.I. Industriales, Universidade de Vigo, Lagoas-Marcosende s/n, 36200 Vigo, Spain

bleon@uvigo.es

Abstract. Lasers can be an effective tool for tailoring the surface of medical implants. Laser irradiation can modify the surface wettability, bioactivity and its capacity to absorb proteins. By using appropriate energies and wavelengths, also the topographical features at macro, micro and nano level can be shaped in order to adapt to cells, extracellular matrices and orientation of ligand molecules. Pulsed laser deposition can produce nanometer thick, dense and well adhering CaP coatings with extremely fine control of chemistry and crystallinity. No further thermal annealing is needed. In-vitro and in-vivo experiments with different cells and animals models have demonstrated similar or better osseointegration of laser deposited coatings compared to the commercial available plasma sprayed ones. Ultraviolet lasers can successfully chemically functionalize the surface of implants, and femtosecond laser can drill polymer plates or meshes for tissue engineering applications.

1. Introduction. The surface of medical bone implants.

A 2003 estimation of the orthopaedic worldwide market amounted 14.2 B€, while the global dental market was estimated to be 1.200 M€ in 2004. Growth rates for both were predicted to be 13-18% annually. In orthopaedics this means worldwide some 2 million hips and 1 million knee implants, with around 300.000 revisions per year. A recent 2010 study [1] confirmed the optimistic predictions of the last decade by estimating presently the US implantable medical device market in $33 billion and in spite of the present economic crisis still they forecast an 8.3% annually growth through 2014.

When implanted, any medical bone substituting device interacts with the body in the initial stages primarily through its surface. Therefore it is of utmost importance to control, design and engineer the surface of implants in order to obtain a fast osseointegration with a long lasting fixation [2]. Thus, all steps in the series of events taking place after implantation are to be tackled for optimization of the correct body response: the protein adsorption into the surface, cell arrival and adhesion to the surface, cell differentiation, proliferation and communication, the secretion of extra cellular matrix (ECM) and later vascularization of the new tissue.

Among the characteristics of an implant surface that will determine its performance in the body, the surface energy determines the wettability of the surface, of great importance since implants are first of all in contact with the water based body fluids. Moreover, depending on the chemical composition of the surface, some glasses and ceramics show bioactivity [3], a time dependent kinetic modification of the surface upon implantation, stimulating surface reactions and leading to bone bonding. Protein adsorption takes place in a time scale from less than a second to minutes for the formation of a monolayer of adsorbed proteins well before cells arrive at the surface. These proteins either promote or prevent cell adhesion.

Moreover, the size and geometry of features must be congruent with the cell morphology, its size, its anchorage focal points and the later formation of extra cellular matrix that will fill the spaces and serve as attachment structures for proteins and cells. In addition, roughness determines the surface area offered for interaction with the environment. Recently nano-roughness has been found to affect the orientation of ligand binding molecules, and therefore modulating their efficiency.

2. Laser processing of surfaces

Lasers can be an effective tool in tailoring the previous mentioned properties of an implant surface for achieving best in-vivo performance, such as laser oxidation of the surface of titanium and its alloys, macro- and micro-structuration of metallic and polymeric implant surfaces [4], laser drilling of polymer plates or meshes for tissue engineering applications using femto-second lasers [5], pulsed laser deposition (PLD) of bioactive very thin coatings [6] or grafting of amine groups on the surface of polymers [7].

2.1 Bioactive coatings

Different research groups have optimized the deposition conditions for production of thin calcium phosphate coatings [6]. Great effort has also been dedicated to understand the processes occurring as the material is transported from target to substrate, in search of scientific bases for the optimization of the coatings beyond empirical work. It has been demonstrated that pulsed laser deposition can produce very thin, dense, well adhering coatings, with extraordinary controlled chemistry and crystallinity. Compared to other deposition techniques, PLD offers several advantages:

- Congruent deposition may be obtained. Different calcium phosphate phases with and without diverse substitutions, and morphologies can be deposited, so that the degree of resorption may be adapted to a specific medical application. Coatings with graded composition, or graded crystallinity can readily be produced.
- No post-deposition thermal annealing is needed, thus the coating-metal interface remains strong adherent.
- Film thickness can be very well controlled, simply by turning the laser on and off.
- The possibility of coating is not limited to metal substrates, but polymer surfaces can also be coated, via either a pre-treatment or modifications of the PLD technique.

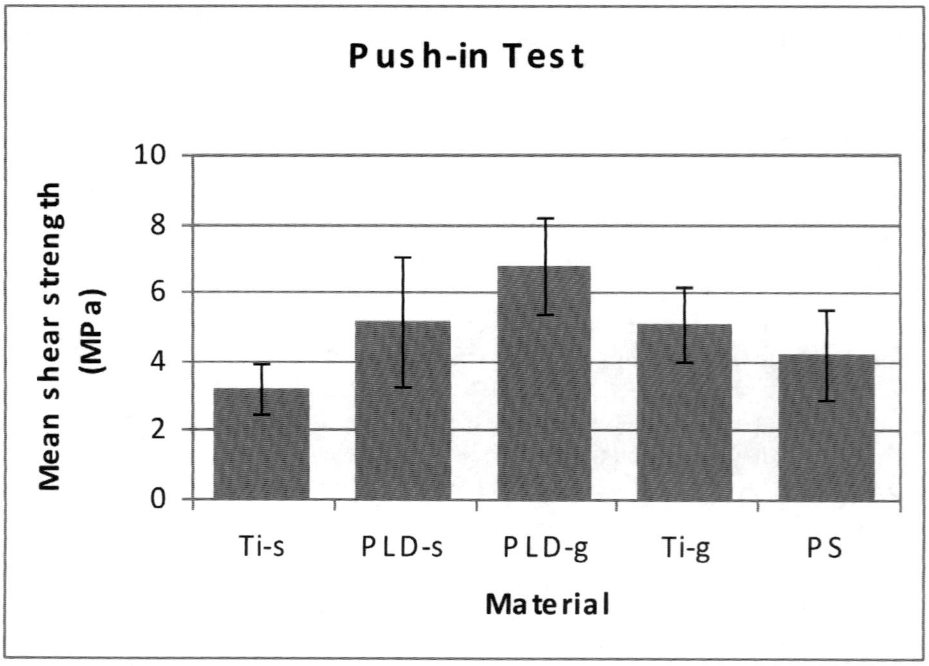

Figure 1. Mean shear strength of titanium cylinders with different surface treatments after implantation in dog jaw (3 months).

In-vitro and in-vivo testing have verified similar or better osseointegration of the pulsed laser deposited hydroxylapatite (HA) films than the commercial available plasma sprayed coatings, but with improved adhesion properties, without risk of delamination or detachment of the coating, as can be the case with the commercial plasma sprayed technique. Fig. 1 depicts the mean shear strength obtained in a push-in test for cylindrical samples that were implanted for three months in dog jaw. Five types of surfaces were tested: smooth as-machined titanium (Ti-s), grit-blasted titanium (Ti-g), plasma sprayed coated HA (PS), pulsed laser deposited HA on grit-blasted titanium (PLD-g) and PLD coatings on smooth titanium (PLD-s).

Surface roughness lead to higher shear strength, due to mechanical interlocking. The PS coating did not improve bonding compared to grit blasted Titanium, probably due to internal fractures within the thick coating when subjected to load. PLD improves bone-bonding (both on grit-blasted and smooth surfaces) in agreement with histomorphometrical findings, whereby more bone apposition was found on the PLD coated samples.

Therefore, the technique is mature for an industrial scale-up on the one hand and the start of clinical tests with real implants on the other hand. Commercial dental implants and orthopaedical spine screws have been coated with HA using ArF laser- demonstrating the uniformity of the coating thickness around the corners and sidewalls of implant devices.

2.2 Chemical functionalization

193-nm excimer laser have been applied to chemically grafting of ammonia on poly(ethylene terephthalate (PET) films. Surface chemical patterning was performed by irradiating PET film in a vacuum chamber filled with ammonia at different fluxes. While the hydrophilicity of the surface increased after patterning and a minimum water contact angle was obtained at the gas flux of 20 ml/min, roughness remained unchanged. TOF-SIMS ion mapping was used to identify the amine containing fragments corroborating that amino grafting mainly happened inside the laser irradiation area of the PET surface. A hypothesized radical reaction mechanism proposes that the collision between the photolytic decomposed radicals from ammonia and those on the PET surface caused by the incident laser provokes the grafting of amino groups.

3. Conclusions

Lasers can be an effective tool for tailoring the surface of medical implants for its variety in wavelengths, pulse lengths and pulse energies. Depending on these parameters, lasers can remove or topographically shape the surface without modifying its chemistry, or chemically vary the surface composition, allowing its chemical functionalization and patterning, locally or in great areas, without compromising its topography.

Acknowledgements

Partial support of EU CRAFT BRE2.CT94.1533, Xunta de Galicia 2006/12, and MAT2003-09277-C03-01 Science & Innovation Ministry, Spain, is acknowledged.

References

[1] Freedonia Focus on Implantable Medical Devices 2010, *Freedonia Report* **FF40032**.
[2] Ratner BD, Hoffman AS, Schoen FJ, and Lemons JE 1996 *Biomaterials Science. An Introduction to Materials in Medicine* (San Diego: Academic Press).
[3] Hench LL and Wilson J 1993, *An Introduction to Bioceramics* (Singapore: World Scientific)
[4] Paz MD, Chiussi S, González P, Serra J, León B, Alava JI, Güemes I, Sánchez-Margallo FM 2008 *Key Eng Mat* **361-363 I** 625.

[5] Martins AA, Wu G, Pinho E, Rebollar E, Chiussi S, Reis R, León B, Neves N, 2010 *J. Tiss. Eng.& Regen. Med.* **accepted**
[6] León B 2009 *Thin Calcium Phosphate Coatings for Medical Implants* (New York: Springer)
[7] Wu G, Paz MD, Chiussi S, Serra J, González P, Wang YJ, and León B 2009 *J.Mater.Sci.: Mater.m.* **20** 597.

Biofunctionalization of titanium surfaces for osseintegration process improvement

P Sevilla[1,2,*], M Godoy[2], E Salvagni[1,2], D Rodríguez[1,2], F J Gil[2]
[1]Nanoengineering Research Center(CRnE), Technical University of Catalonia. Av. Diagonal 647. Barcelona, Spain

[2]Biomaterials, Biomechanics and Tissue Engineering Group (BIBITE), UPC.

*E-mail: pablo.sevilla@upc.edu

Abstract: This study aims to improve the osseointegration of titanium implants through surface immobilization of peptides that induce a beneficial biological response. This was carried out biofunctionalizating titanium surfaces by silanization and subsequent covalent binding of a peptide with a sequence that promotes cell adhesion. *Objective:* The development of a new technique of immobilization of oligopeptides on the surface of titanium by using 3-chloropropyltrietoxisilane (CPTES) as bonding agent between the surface of titanium and the peptide. *Materials and methods:* A physicochemical characterization of the surfaces through the techniques of XPS, ToF-SIMS and contact angle was performed. Also cell adhesion studies have been conducted to evaluate in vitro biological response. *Results:* Through the process of silanization the titanium surface is completely covered with CPTES, which allows the subsequent accession of oligopeptides. The cell adhesion results show a higher cell adhesion and cell extension on biofunctionalized samples. *Conclusions:* We developed a system of covalent binding of oligopeptides on titanium surfaces that can modify the biological response of the attached cells.

Keywords: biofunctionalization, Titanium, Osseointegration.

1. Introduction

Every year millions of metallic implants are implanted in patients, at a cost of billions of dollars in the whole world, to restore bodily functions that have degenerated due to old age, disease or trauma [1, 2]. When these elements are brought into contact with the physiological environment, it starts a chain of events that are dominated by protein adsorption and subsequent adhesion of platelets and macrophages, inducing inflammation of the affected area. This nonspecific response to the implant is considered not optimal for a good performance of the implant in the organism [3]. The cascade of reactions developed by the human immune system depends on the physicochemical properties of the surface. The acceptance of an implant by the human body is given by the durability, integrity and surface characteristics of the implant, as

well as chemical and biological reactions which occur within the tissue-implant interface [4, 5]. Several approaches to develop a new generation of metalllic implants and medical devices are currently being studied. These new devices are designed to interact with the host body in a controlled and predictable way to replace diseased or damaged tissues. Some of these proposals lie in the incorporation of biomacromolecules, such as peptides, proteins, enzymes, lipids or DNA plasmids, on the surface of the implant material [6-11]. Of these proposals, the use of peptides to get a direct connection between the implant and cell surface receptors has gained wide acceptance and seems to have great potential due to the ease of modification of a broad range of materials [12]. Often, the peptides are designed to mimic small domains found in proteins present in the extracellular matrix of living tissue to promote cell activity. Modifying the surface prior to implantation, the material ensures an interaction with specific cell types through biomolecular recognition processes, thus avoiding nonspecific interactions between the implant and the surrounding tissue. This level of specificity in cell interaction and growth of native extracellular matrix significantly reduces recovery time after implantation. Accelerating the healing process, the patient may use faster the implant in service. This strategy has received in recent years, the name of biomimetics [14]: "manufacture materials that mimic the composition, structure or process of synthesis of natural materials." More specifically the surfaces modified with these processes are called biomimetic surfaces.
In this work we have developed a new method for binding short peptides on the surface of titanium to obtain biomimetic surfaces based on the use of peptides capable of controlling cell activity both in vitro and in vivo.

2 . Materials and Methods

Starting from commercially pure grade two titanium bars, 10 mm diameter and 2 mm thick disks were machined. The titanium discs were trimmed with silicon carbide paper and polished with colloidal silica to obtain a surface roughness Ra between 10 and 15 nm. Subsequently, the discs were washed three times with the following solvents: cyclohexane, isopropanol, distillied water and acetone, sonicated for 15 minutes each and were finally dried with nitrogen. In order to activate titanium surface for further silane binding the titanium discs were treated with plasma cleaning (Plasma Cleaner, Sterilizer PDC-002, Harrick Scientific Corporation, USA) for 5 minutes. Immediately after the disks were silanized by soaking in a solution of 0.85 ml of pentane, 0.075 ml of Diisopropylethylamine (DEA) and 0.15 ml of 3-chloropropyl triethoxysilane (CPTES) per disk. To avoid the incorporation of water into the solution, the silanization was carried out in an inert atmosphere of nitrogen for 1 hour with gentle shaking. After silanization, titanium disks were washed in a series of ethanol, isopropanol, acetone and distilled water and dried with nitrogen. Then titanium disks were immersed in an aqueous solution at pH 11.0 with 100 μmol of dissolved peptide per sample for 18 h, and finally washed in ethanol, isopropanol, distilled water and acetone and dried with nitrogen. The peptides used for the study were provided by GenScript USA Inc. The peptide amino acid sequences selected were: NH2-KGRGDS-COOH, NH2-KGRDGS-COOH and NH2-KGGGRGDS-COOH. The RGD sequence has the potential of cell adhesion [13]. To contrast the information obtained, clean titanium control samples and non silanized samples exposed to the different peptide solutions were also studied.

The physicochemical characterization of the obtained surfaces was carried out with X-ray Photoelectron Spectroscopy (XPS), Time of Flight Secondary Ion Mass Spectroscopy (ToF-SIMS), fluorescence microscopy and contact angle. The XPS analysis was performed using a spectrometer NanoESCA (Omicron Nanotechnology, Germany) imposing a step energy of 150 eV. To offset the effects of residual carbon peak the spectra was referenced to 284.8 eV. The ToF-SIMS analysis was performed with a ToF SIMS IV (Ion Tof, Germany) using cesium ion as a surface ionizer. The contact angle measurements were performed using a Goniometer OCA 15 + (Dataphysics, Germany). For each measure the sessile drop method with 300 μl of distilled water for each measure was used.

To observe the cellular response, samples were biofunctionalized with the three different peptides previously mentioned. Titanium samples with peptides adsorbed on the surface were prepared for comparison purposes. The cells selected for adhesion testing were human osteoblasts MG-63 line. Prior to exposure of cells,titanium samples were washed in ethanol for 1 h and then were blocked with bovine serum albumin 6% in PBS for 1h. Osteoblasts were exposed to titanium surfaces for 3 h in serum free medium and then were fixed and stained with phalloidin and Hoetsch (Sigma Aldrich, USA) for 45 minutes at 37 ° C to reveal its and actin and nuclei respectively.

3. Results

XPS: Table 1 shows the chemical composition present on the surface of the samples. The presence of chlorine on silanized surfaces is an indicator of the presence of CPTES on the surface. There is also an increase of nitrogen on the samples that have been exposed to a peptide. Since the only molecule that contains nitrogen in the whole system of study is the peptide, this data indicates the presence of the peptide on the surface.

Table 1. Chemical composition of the developed surfaces. Results obtained by means of XPS.

%at	Ti	Ti+CPTES	Ti+CPTES+RGD	Ti+RGD
O1s	49.6	41	35.5	43.9
C1s	26.1	39.1	42.7	34.9
Ti2p	20.8	13.7	12.7	18.8
Si2p	2.2	3.3	4.3	nd
N1s	1.1	0.5	2.5	2.3
Cl2p	0.1	2.4	2.2	nd

ToF-SIMS: Positive ion spectra obtained by ToF-SIMS is shown in figure 1. For the silanized and peptide KGRGDS biofunctionalized samples sharp spikes in values of mass/charge 28, 30, 42, 56, 70, 85, 97 and 111, corresponding to ions $CH2N^+$, $CH4N^+$, $C2H4N^+$, $C3H6N^+$, $C4H8N^+$, $C5H4N^+$ $C7H13N^+$ and $C6H11N^+$ respectively. All these species contain nitrogen, indicating the presence of the peptide KGRGDS on the surface. In the case of the biofunctionalized sample without silane these peaks are also present although with less intensity.

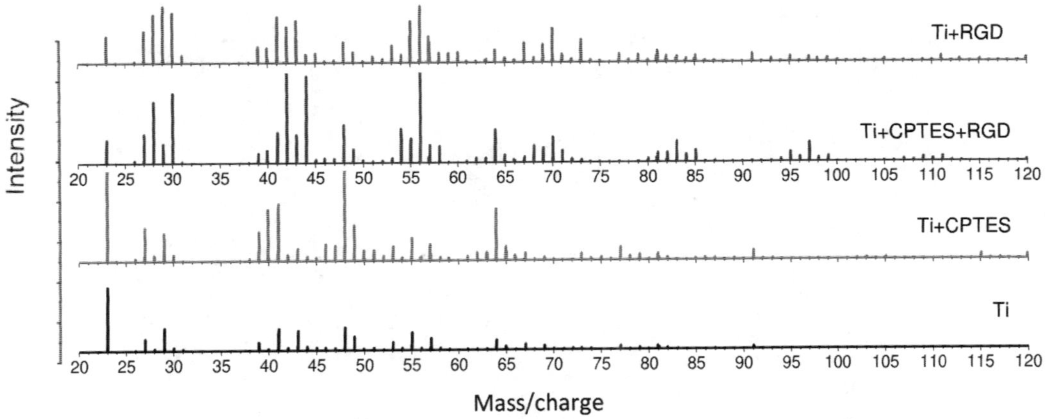

Figure 1: ToF-SIMS spectra obtained on the treated and non treated samples.

Contact angle: The results of contact angle (Fig. 2) show significant differences between silanized and non silanized samples, increasing the contact angle in the first case. The alkyl chain of the CPTES has hydrophobic properties increasing the contact angle of silanized samples as compared to the contact angle of clean titanium. The reduction of the contact angle of the silanized and biofunctionalized samples with the peptide KGRGDS when compared to contact angle of silanized titanium indicates the presence of the peptide on the surface because it is hydrophilic. There were no statistically significant differences between the contact angle of clean titanium and biofunctionalized titanium without CPTES, suggesting that the amount of peptide adsorbed, in the last case, is lower than in the case of silanized titanium with CPTES.

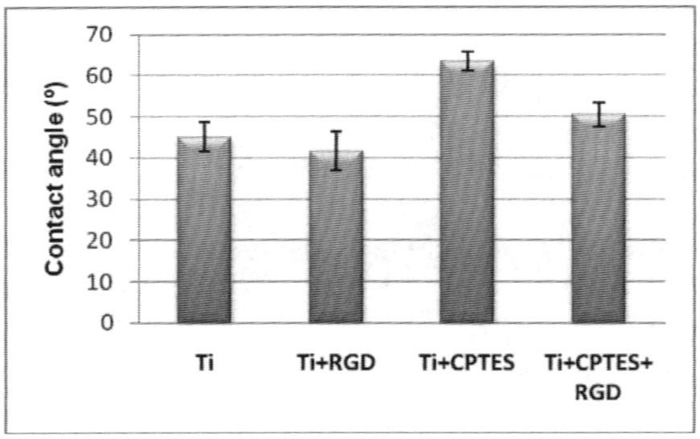

Figure 2. Contact angle results obtained on treated and non treated titanium samples.

Cell adhesion: Cell adhesion results (Fig. 3) showed a higher cell affinity for the biofunctionalized samples with the RGD sequence. Table 2 shows the number of cells per area unit and surface area covered by them. These two values are indicative of cell adhesion on the studied surfaces. Although the greatest number of cells adhered to the surface on the samples biofuntionalized with the peptide KGRGDS there is a greater cell extension on the biofunctionalized sample with the peptide KGGGRGDS.

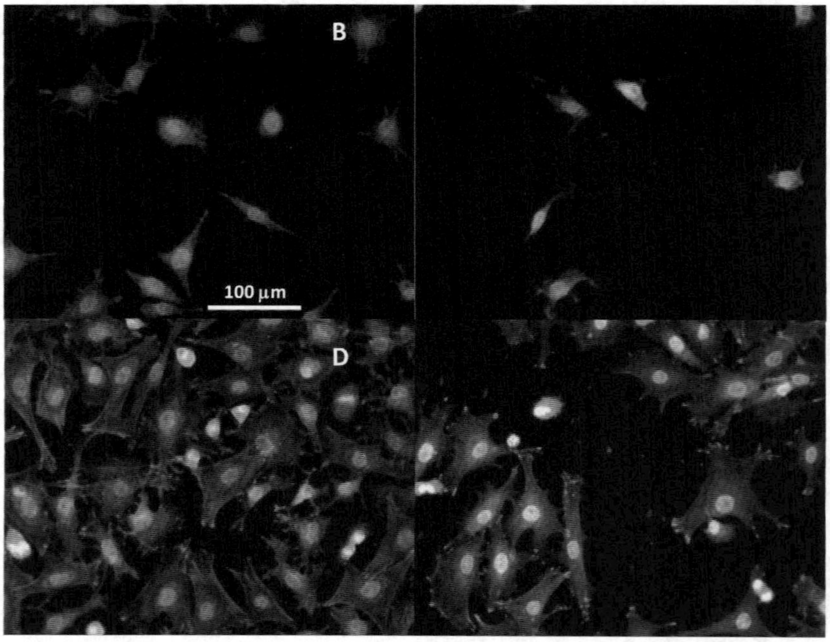

Figure 3. Fluoresecence micrographs of the MG-63 attached to different studied surfaces. A: Clean titanium. B: Ti+CPTES. C: Ti+CPTES+KGRGDS. D: Ti+CPTES+KGGGRGDS.

Table 2. Cell adhesion on the studied samples.

	Cell number/cm^2	**Cell extension (%)**
Clean Ti	657.2 ± 118.8	41.4 ± 10.7
Ti+CPTES	45.0 ± 7.2	16.7 ± 13.1
Ti+KGRGDS	589.7 ± 78.4	34.5 ± 10.1
Ti+KGGGRGDS	625.7 ± 58.7	37.5 ± 11.3
Ti+KGRDGS	270.1 ± 11.7	28.8 ± 12.7
Ti+CPTES+KGRGDS	1960.60± 159.0	48.3 ± 15.5
Ti+CPTES+KGGGRGDS	847.7 ± 114.4	58.2± 20.9

4. Discussion

The XPS results have shown the presence of a CPTES layer on the surface of titanium in silanized samples. The shallow penetration depth of this technique is about 5 nm. As it can be seen comparing the clean titanium sample and the completely biofunctionalized titanium, the presence of titanium element is significantly reduced with the Biofunctionalization treatment but the XPS still detects a significant amount of it. This indicates that the layer formed has a thickness of less than 5 nm, possibly a monolayer with a thickness of about 1 nm is formed [15]. This thickness seems to increase by the addition of the peptide because the amount of titanium detected is even lower. On the other hand, the increase in the percentage of nitrogen in the samples indicates the presence of peptide itself. Moreover, although the ToF-SIMS technique is not quantitative, the results allow inferring that there is a greater amount of peptide on silanized surfaces than in those non-silanized.

It has also been observed that the silanized samples have a more hydrophobic behavior than clean titanium. Although it is not the only factor involved, this can significantly influence the cell adhesion as has been observed in cell adhesion tests. Silanized samples without peptide presented a minimum cell adhesion; it means that cell adhesion is mediated, mainly, by the peptides present on the titanium surfaces. Comparing KGRDGS and KGGGRGDS peptides on the sample surfaces, there were a higher number of cells adhering to the surface in the case of silanized samples and biofunctionalized with KGRGDS peptide. Nevertheless cell extent in the silanized biofunctionalized samples with the peptide KGGGRGDS is somewhat higher. The RGD sequence is recognized by cells through their integrins, mainly $\alpha_5\beta_1$ integrin binding to this sequence and hence to the titanium surface. RDG sequence is a sequence of control which has no cellular adhesion potential. The inclusion of one or three glycines in the peptide can give differences in cell adhesion since it is possible that the peptide KGGGRGDS, having the RGD sequence further away from the surface may allow easier access for cellular recognition. These data could indicate a greater accessibility of the peptides KGGGRGDS for the cells.

5. Conclusion

A new method of immobilization of oligopeptides on titanium surfaces has been developed. The silanization with CPTES before biofunctionalization allows a more selective cell adhesion than in the case of a biofunctionalized titanium surface without prior silanization.

References

[1] Hench LL 1999 Science **208** 826
[2] Agrawall C M 1998 J. Min. Met. Mat. Soc **50** 31
[3] Healy K E, Rezania A and Stile R A 1999 Ann. N.Y. Acad. Sci. **875** 24
[4] Hanker J S and Giammara B L 1988 Science **242** 885
[5] Kasemo, B and Lausmaa J 1988 J Oral Maxillofac. Implants **3** 247

[6] Muller R, Abke J, Schnell E, Scharnweber D, Kujat R, Englert C, Taheri D, Nerlich M and Angele P 2004 Biomaterials **27** 4059
[7] Radler J and Sackmann E 1997 Curr. Opin. Solid St. M. **2** 330
[8] Anseth K S, Benoit D S W and Durney A R 2007 Biomaterials. **28** 66
[9] Tosatti S, Schwartz Z and Campbell C 2004 J. Biomed. Mater. Res. **68A** 458
[10] Tabata Y, Yamada K, Miyamoto S, Nagata I, Kikuchi H, Aoyama I, Tamura M and Ikada I 1998 Biomaterials **19** 807
[11] Hollinger J O, Winn S and Bonadio 2000 Tissue Eng, **6** 341
[12] Dillow A K and Lowman A M 2002 Biomimetic materials and Design Ed. Marcel Dekker Inc.
[13] Verné E, Ferraris S, Vitale-Brovarone C, Spriano S, Bianchi C L, Naldoni A, Morra M and Cassinelli C 2010 Acta Biomat. **6** 2010
[14] Ruoslahti E and Pierschbacher M D 1987 Science **238** 491
[15] Lee M H, Brass D A, Morris R, Composto R J and Ducheyne P 2005 Biomaterials **26** 1721

Journal of Physics: Conference Series **252** (2010) 012010 doi:10.1088/1742-6596/252/1/012010

Evaluation of magnesium alloys with alternative surface finishing for the proliferation and chondro-differentiation of human mesenchymal stem cells

J. Trinidad[1], G. Arruebarrena[1], E. Sáenz de Argandoña[1], G. Ruiz de Eguino[2], A. Infante[2], C. I. Rodríguez[2]

[1] Mondragon Unibertsitatea, Loramendi 4, Mondragón, Spain
[2] Stem Cell Laboratory, Research Unit, Hospital de Cruces, Plaza Cruces s/n, Barakaldo, Spain

E-mail: `jtrinidad@eps.mondragon.edu`

Abstract. Articular cartilage has little capacity for self-repair. As a result, continuous mechanical stress can lead to the degradation of articular cartilage, culminating in progressive damage and joint degeneration. Tissue engineering has arisen as a promising therapeutic approach to cartilage repair. Magnesium alloys are one of the most important metallic biomaterials emerging in this area due to their biocompatibility, bio-absorbability and especially to their mechanical properties. These properties make magnesium alloys a promising biomaterial in the regeneration of cartilage tissue. *Objective.* This study was undertaken to analyze the influence of surface characteristics of magnesium alloys in the adhesion, proliferation and differentiation of human mesenchymal stem cells (MSCs). *Methods.* Two commercial magnesium alloys (AZ31B and ZM21) were subjected to different treatments in order to obtain four different surfaces in each alloy. Human MSCs were seeded into the magnesium alloys and analyzed for their proliferation and chondrogenesis differentiation ability. *Results.* Human MSCs showed a greater proliferation and chondro-differentiation when cultured in the ZM21 magnesium alloy with a surface finishing of fine sanding, polishing, and etching.

1. Introduction

Damage to articular cartilage is of great clinical consequence because it has a poor regenerative capacity due to the lack of blood supply and the inability of native chondrocytes to proliferate and participate in regeneration at the injury site [1, 2].

Tissue engineering combines the principles of engineering and biology to design and fabricate constructs to promote and expedite the repair and regeneration of damaged tissue. The underlying principle of tissue engineering involves the utilization of biocompatible and mechanically suitable scaffolds, combined with an appropriate source of cells. These components, when combined, form a tissue-engineered construct which can function as the tissue replacement material, and, in principle, facilitate a faster rate of tissue repair [1–3].

Current focal cartilage lesion treatments are based on surgical procedures like abrasion arthroplasty, microfracture, mosaicplasty, osteochondral allografts and autologous chondrocyte implantation (ACI, MACI) [4]. These techniques are based on the generation of fibrous cartilage that covers the defect. This fibrocartilage is histologically different from articular cartilage, containing a higher proportion of type I collagen fibers with less glycosaminoglycans. Hence, the fibrous cartilage has less than optimal physical and biochemical characteristics compared to articular cartilage. Furthermore, the integration rate of the new extracellular matrix with the injured adjacent cartilage is low due to the anti-adhesive properties of the new matrix. Moreover, with these techniques long term clinical results indicate that degenerative joint diseases such as osteoarthritis may appear [4, 5].

Mesenchymal stem cells (MSCs) are a promising cell source for the engineering of cartilage tissue, due to their intrinsic capacity to differentiate to chondrocytes [6]. To improve the results in the healing of articular cartilage, as seen in Ohba's review [7], research groups have generated articular cartilage through the *in vitro* differentiation of mesenchymal stem cells (MSCs) on scaffolds.

Successful cell-based cartilage tissue engineering depends not only on the selection of an appropriate biocompatible scaffolding material, but also on materials with suitable mechanical properties that result in a desirable clinical outcome when combined with cells.

Magnesium (Mg) and its alloys are new biomaterials which have been studied recently for tissue engineering [8–15]. The advantage of Mg compared to other biomaterials traditionally used in tissue engineering, such as polymers, injectable gels and ceramics, are its mechanical properties. Magnesium has a higher mechanical strength than polymers and gels, and greater fracture toughness than ceramics. In addition, Mg is an essential component in human metabolism [9]. It is the fourth most abundant cation in the body and the second most important after potassium in the intracellular environment [16]. The Mg of the human body appears mainly in the skeleton, being an essential component for bone growth and maturation [17, 18]. A Mg deficit may cause disease in the cartilage [19], and moreover Mg ions can increase the rate of proliferation of chondrocytes in *in vitro* culture [20]. Finally, taking into account its degradation in the human body, the cations generated due to Mg corrosion are efficiently regulated by the organism [14]. These characteristics make Mg a promising material for use in musculoskeletal tissue regeneration. However, the main problem of Mg is its high rate of degradation in body fluids [8–11, 15]. The rapid degradation of Mg could generate subcutaneous hydrogen bubbles that can damage surrounding tissues.

In this study, human MSC are induced to proliferate and to differentiate into chondrocytes on two commercial Mg alloys. The influence of material and surface finishing is evaluated based on the cellular proliferation potential observed and the quality of generated tissue.

2. Experimental procedure

2.1. Choice of the magnesium alloys

The magnesium alloys evaluated in this study are AZ31B (nominal 3%wt. Al and 1%wt. Zn) and ZM21 (nominal 2%wt. Zn and 1%wt. Mn). Aluminium is an alloying element that increases the corrosion resistance of magnesium. In fact, magnesium alloys combining with aluminium have been commonly used, especially aluminium-zinc series, in several studies [10–13, 15], despite Al being a toxic element [20]. AZ31B was chosen because it contains low percentage of Al ($\sim 3\%$), decreasing the toxicity caused by Al. The second alloy, ZM21, has been chosen due to the corrosion resistance that magnesium alloys with similar composition have shown in body

Table 1. Surface finishing treatment.

Surface	Treatment
Surf. 1	1000 grit SiC paper SiC
Surf. 2	1000 grit SiC paper + 4000 grit SiC paper
Surf. 3	1000 grit SiC paper + 4000 grit SiC paper + polishing (Al_2O_3 0.05 μm)
Surf. 4	1000 grit SiC paper + 4000 grit SiC paper + polishing (Al_2O_3 0.05 μm) + etching (HNO_3 69.5 %)

fluids [21].

2.2. Development of magnesium samples

20x20x2 mm samples were prepared. They were processed to be divisible into 4 pieces of 10x10 mm for various analyses (cellular adhesion and proliferation, chondrogenesis and material degradation). In order to determine the influence of surface finishing on cellular adhesion and proliferation, four different surfaces were evaluated. These surfaces are described in Table 1.

After modifying the surfaces of the samples, they were subjected to a sterilization process with ethylene oxide to avoid possible alterations due to thermosensitivity of the material [22].

2.3. Mesenchymal stem cells

The cells chosen for culture on magnesium alloys are primary cultures of human mesenchymal stem cells (MSCs) obtained from bone marrow. The MSCs are a type of adult stem cell with a high rate of proliferation *in vitro* during the first passages. Furthermore, they are able to differentiate under appropriate culture conditions into mesodermal cell lineages (adipocytes, osteocytes and chondrocytes) [6]. In the present work the proliferation and chondrogenesis potential of the MSCs on magnesium alloys has been studied.

Human bone marrow-derived MSCs (BM-MSCs) were obtained from Inbiobank, a node of the National Cell Line Bank. These cells were harvested from cadaveric donors upon the informed consent of close relatives under the supervision of the Spanish National Transplant Organization (ONT). After expansion, MSCs displayed a typical CD29+, CD73+ (SH3 and SH4), CD105+ (SH2), CD166+, CD34-, CD45- and CD31- phenotype. In the presence of specific differentiation factors, and they were able to differentiate into adipocytes, chondrocytes and osteocytes (not shown).

2.4. MSCs proliferation

MSCs were spread at high density on Mg alloys and were maintained in culture for two weeks in a controlled atmosphere with 5% CO_2 at 37 °C. The culture medium for cell proliferation was enriched with proteins, amino acids and antibiotics to avoid possible contamination. The components are detailed in Table 2.

For the cell proliferation analysis on Mg alloys, the cells were stained with Hoechst. This is a fluorescent stain that labels double-stranded DNA concentrated in the cell nucleus, allowing

Table 2. Components of the proliferation medium and differentiation medium.

Medium type	Medium components
Proliferation	DMEM L-Glutamine Penicillin/Streptomycin
Chondrogenesis	Basal Medium Dexamethasone Sodium piruvate L-Glutamine Proline Gentamicin/Amphotericin Ascorbic acid TGF-β

the identification of cells on metallic alloys using fluorescence microscopy. The presence of this stain does not affect cell viability.

2.5. Differentiation into chondrocytes: chondrogenesis

Once the cells proliferated on the alloys, chondrogenesis was induced. Cells were cultured in differentiation medium supplemented with recombinant growth factor TGF-β, described in Table 2. The cells were maintained for 6 weeks under these conditions, replacing the culture medium every 2 or 3 days.

3. Results

During the cell culture, the generation of gas bubbles was observed in Mg alloys (Figure 1). These bubbles were generated due to Mg corrosion in the culture medium. The Mg corrosion is described by the equation (1) [8]:

$$Mg(s) + 2H_2O \rightarrow Mg(OH)_2 + H_2(g) \quad (1)$$

According to this equation, the gas generated in the culture medium is hydrogen.

The generation of hydrogen was variable depending on the materials and surface finishing. The ZM21 alloy, generated less hydrogen, and had a greater corrosion resistance in physiological medium than the AZ31B alloy. Regarding the surface finishing, Surf. 1 presented less corrosion resistance in both materials. The corrosion resistance of Surf. 2, Surf. 3 and Surf. 4 did not differ significantly. In the same way, a pH increment of the culture medium was produced due to the generation of hydroxides (equation (1)).

Analysis of the corrosion process on the magnesium alloys indicated that, corrosion products were observed in the culture medium after several days of proliferation (Figure 2a). Despite this observation, the cells were able to adhere to the samples, as shown in Figure 2b.

Figure 1. Detail of the generated hydrogen bubbles due to the corrosion of the AZ31B Surf. 1 in the culture medium.

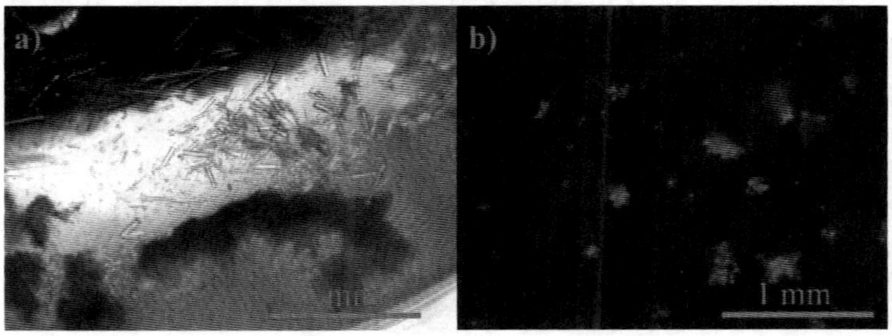

Figure 2. Cell proliferation on ZM21 Surf. 1: a) Corrosion products b) Groups of cells by fluorescence microscopy

After two weeks in proliferation medium, cell confluence was great enough for the formation of a monolayer of undifferentiated cells. This cell layer covered the sample partially with significant extensions of tissue. Figure 3 shows one of the best examples of adhesion and cell proliferation (ZAM21 Surf. 4).

Comparing the two alloys, cell proliferation was lower, or nonexistent, in AZ31B alloy compared to the ZM21 alloy, as shown in Figure 4.

Taking into account the different surface morphologies of the ZM21 alloy, a greater cell proliferation rate was observed on surfaces 2, 3 and 4 (Figure 5). Surfaces 2 and 4 showed a more uniform cell distribution. The polishing of Surf. 3 apparently resulted in compromised cell adhesion and proliferation.

After the chondrogenesis process, the composition of the secreted matrix by differentiated cells was analyzed. For that purpose the sulfated glycosaminoglycans of the matrix were quantified, because these components are the principal constituents of the extracellular matrix of cartilage. In this analysis sulfated glycosaminoglycans were found. Preliminary results indicated a greater potential for the ZM21 alloy compared to the AZ31B alloy in regenerative medicine requiring

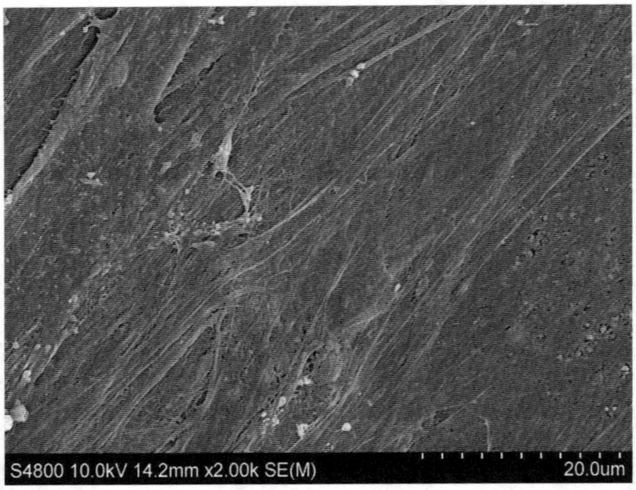

Figure 3. Detail of cell monolayer on ZM21 Surf. 4. Optimal cell proliferation region observed by SEM.

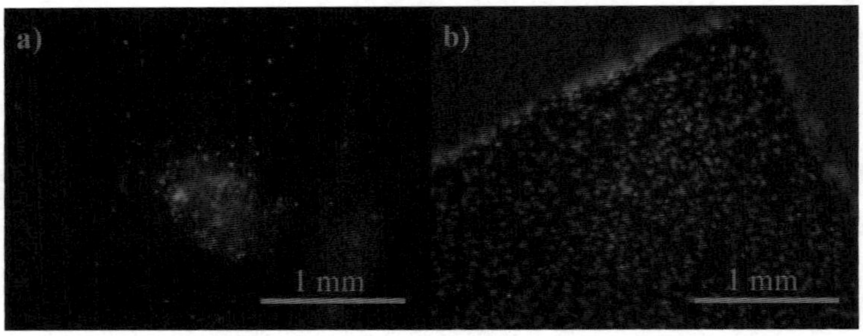

Figure 4. Hoechst staining after cell proliferation: a) AZ31B Surf. 2 b) ZM21 Surf. 2

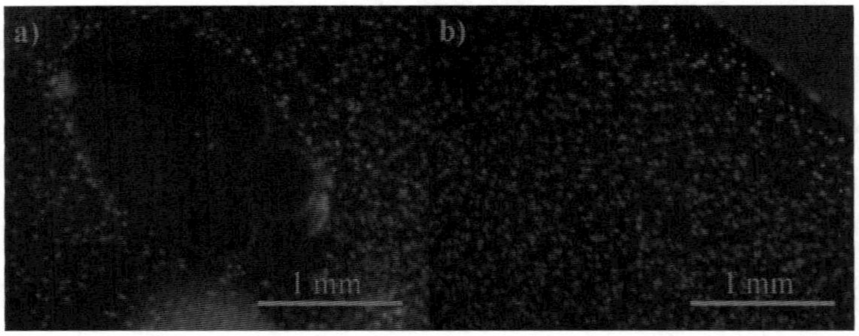

Figure 5. Hoechst staining after cell proliferation: a) ZM21 Surf. 3 b) ZM21 Surf. 4

the generation of cartilage.

4. Conclusions

The preliminary results of this study indicate that the ZM21 alloy has a lower corrosion rate than AZ31B in simulated physiological fluids. The ZM21 alloy also favors cell adhesion and subsequent cell proliferation of mesenchymal stem cells on it; while the AZ31B alloy has a lower cell adhesion and proliferation rate.

Surface finishing, with fine sanding (Surf. 2), polishing (Surf. 3) and etching (Surf. 4) gave a homogeneous distribution of cells on the alloys. However, the polishing (Surf. 3) makes adhesion and proliferation of mesenchymal stem cells more difficult.

Subsequent identification of sulfated glycosaminoglycans in the samples confirmed the chondrogenisis process on magnesium alloys.

Acknowledgements

This work was funded by the Department of Industry, Innovation, Trade and Tourism of the Basque Government through the SAIOTEK program.

References

[1] Tae S K, Lee S H, Park J S and Im G I 2006 *Biomedical Materials* **1** 63–71
[2] Chen G, Liu D, Tadokoro M, Hirochika R, Ohgushi H, Tanaka J and Tateishi T 2004 *Biochemical and Biophysical Research Communications* **322** 50–55
[3] Hardingham T, Tew S and Murdoch A 2002 *Arthritis Research* **4 Suppl 3** S63–S68
[4] Magnussen R A, Dunn W R, Carey J L and Spindler K P 2008 *Clinical Orthopaedics and Related Research* **466** 952–962
[5] Glass G G 2006 *Dis Mon* **52** 343–362
[6] Dominici M, Le-Blanc K, Mueller I, Slaper-Cortenbach I, Mariniand F, Krause D, Deans R, Keating A, Prockop D and Horwitz E 2006 *Cytotherapy* **8** 315–317
[7] Ohba S, Yano F and Chung U 2009 *IBMS BoneKey* **6** 405–419
[8] Zeng R, Dietzel W, Witte F, Hort N and Blawert C 2008 *Advanced Engineering Materials* **10** B3–B14
[9] Staiger M P, Pietak A M, Huadmai J and Dias G 2006 *Biomaterials* **27** 1728–1734
[10] Witte F, Ulrich H, Rudert M and Willbold E 2007 *Journal of Biomedical Materials Research - Part A* **81** 748–756
[11] Witte F, Ulrich H, Palm C and Willbold E 2007 *Journal of Biomedical Materials Research - Part A* **81** 757–765
[12] Krause A, von der Hoh N, Bormann D, Krause C, Bach F W, Windhagen H and Meyer-Lindenberg A 2010 *Journal of Materials Science* **45** 624–632
[13] Gu X, Zheng Y, Cheng Y, Zhong S and Xi T 2009 *Biomaterials* **30** 484–498
[14] Li Z, Gu X, Lou S and Zheng Y 2008 *Biomaterials* **29** 1329–1344
[15] Pietak A, Mahoney P, Dias G J and Staiger M P 2008 *Journal of Materials Science: Materials in Medicine* **19** 407–415
[16] Gums J G 2004 *American Journal of Health-System Pharmacy* **61** 1569–1576
[17] Aranda P, Planells E and Llopis J 2000 *Ars Pharmaceutica* **41** 91–100
[18] Iannello S and Belfiore F 2001 *Panminerva Med,* **43** 177–209
[19] Stahlmann R, Frster C, Shakibaei M, Vormann, Gnther T and Merker H J 1995 *Antimicrobial Agents and Chemotherapy* **39** 2013–2018
[20] El-Rahman S S A 2003 *Pharm Res* **47** 189–194
[21] Song G 2007 *Corrosion Science* **49** 1696–1701
[22] Silvestre C, Fagoaga L, Garcianda M J, Lanzeta I, Mateo M C and Zapata M C 2000 *ANALES Sis San Navarra* **23 (Suppl. 2)** 95–103

Journal of Physics: Conference Series **252** (2010) 012011 doi:10.1088/1742-6596/252/1/012011

Bacterial adherence to anodized titanium alloy

C Pérez-Jorge Peremarch[1], R Pérez Tanoira[1], MA Arenas[2], E Matykina[2], A Conde[2], JJ de Damborenea[2], E Gómez Barrena [1], J Esteban[1]

1 IIS-Fundación Jiménez Díaz Hospital, Madrid, Spain.
2 Centro Nacional de Investigaciones Metalúrgicas CENIM/CSIC. Madrid, Spain

E-mail: cperemarch@fjd.es

Abstract. The aim of this study was to evaluate *Staphylococcus* sp adhesion to modified surfaces of anodized titanium alloy (Ti-6Al-4V). Surface modification involved generation of fluoride-containing titanium oxide nanotube films. Specimens of Ti-6Al-4V alloy 6-4 ELI-grade 23- meets the requirements of ASTM F136 2002A (AMS 2631B class A1) were anodized in a mixture of sulphuric/hydrofluoric acid at 20 V for 5 and 60 min to form a 100 nm-thick porous film of 20 nm pore diameter and 230 nm-thick nanotube films of 100 nm in diameter. The amount of fluorine in the oxide films was of 6% and of 4%, respectively. Collection strains and six clinical strains each of *Staphylococcus aureus* and *Staphylococcus epidermidis* were studied. The adherence study was performed using a previously published protocol by Kinnari et al. The experiments were performed in triplicates. As a result, lower adherence was detected for collection strains in modified materials than in unmodified controls. Differences between clinical strains were detected for both species ($p<0.0001$, Kruskal-Wallis test), although global data showed similar results to that of collection strains ($p<0.0001$, Kruskal-Wallis test). Adherence of bacteria to modified surfaces was decreased for both species. The results also reflect a difference in the adherence between *S. aureus* and *S. epidermidis* to the modified material. As a conclusion, not only we were able to confirm the decrease of adherence in the modified surface, but also the need to test multiple clinical strains to obtain more realistic microbiological results due to intraspecies differences.

1. Introduction

The use of prosthetic joints have been a great advance in orthopaedic surgery. The number of hip and knee arthroplasties has been constantly increasing, with nearly 800.000 of such procedures performed in the United States of America during 2006 [1].

Complications appear in almost 10% of the cases, of which infections play a major role due to significant clinical impairment which entails a great cost to the public health system. Even though only a small percentage of the arthroplasties end in infection, between 0.8 % and 1.9% for knee and between 0.3% and 1.7% for hip, the severity of the problem and the long and difficult management of the affected patients is of great relevance [1].

Causal microorganisms includes a broad spectrum of species, and it depends on the type and location of the prosthesis, elapsed time after the surgery and risk factors of the patient.

Staphylococcus (*Staphylococcus aureus* and coagulase-negative *Staphylococcus* species) accounts for more than half of the prosthetic-hip and prosthetic-knee infection cases, and *Staphylococcus epidermis* and *S. aureus* have been isolated from infected biomaterial surfaces in several studies [1-2]. While *S. aureus* is a known virulent pathogen, *S. epidermidis* is usually a nonpathogenic, commensal, human skin saprophyte, and has emerged as a common pathogen in biomaterial-related infections [2].

Environment, bacterial characteristics, material surfaces, and serum or tissue proteins are the main factors influencing biomaterial-related infections [3-4]. Interactions between bacteria or tissue cells and substratum surface depend greatly on the surface and near surface atomic structure and composition of biomaterials. Biomaterial surfaces may be modified in order to improve compatibility and tissue integration and to resist microbial colonization in the "race for the surface" [2]. This concept suggests that there is a race between tissue cell integration and bacterial adhesion to that same surface. If the race is won by bacteria, the complex process of bacterial adhesion begins [2]. This is described in a two phase phenomenon, which includes an initial instantaneous and reversible physical phase and an irreversible molecular and cellular phase [3].

Titanium implants are usually employed in many biomedical applications such as orthopaedic implants. There are many surface coatings, such as adhesion-resistant coatings and those containing antimicrobial agents. Fluorine is an inorganic antimicrobial agent that has been introduced into titanium, but it has not yet been studied in depth [4]. The purpose of this work is to evaluate the bacteria adhesion after surface modification of Ti-6Al-4V to increase the stability of the surface by means of Fluoride incorporation.

2. Materials

2.1 Surfaces treatment of Ti-6Al-4V.

A 18 mm diameter rod of Ti-6Al-4V alloy ELI grade, according to the standard ASTM F136-02, supplied by SURGIVAL was cut into 2 mm thick disk specimens, ground through successive grades of SiC paper to 1200 grade. The specimens were then chemically polished in a mixture of $HF:HNO_3:H_2O$ with volume ratio 1:4:5 for 5 min at room temperature under continuous agitation.

Nanotubes were formed in a two-electrode cell by anodizing the specimens at 20 V in a electrolyte containing H_2SO_4 and HF for 5 and 60 min according to the patent described elsewhere [5]. Platinum mesh was used as a cathode. Voltage-time and current-time responses of anodizing were acquired at 0.1 Hz sampling rate using a zero-resistance ammeter.

Electrochemical measurements were performed using Gamry Reference 600 potentiostat and a conventional three-electrode cell, comprising a tested specimen with an exposed area of 0.5 cm2 as a working electrode, silver chloride reference electrode and a platinum counter electrode. All the tests were carried out in a naturally aerated phosphate buffer solution (NaCl 8.77 g/l, Na_2HPO_4 1.28 g/l, KH_2PO_4 1.36 g/l, pH 7.2-7.4) at 37oC. Cyclic potentiodynamic polarisation tests were carried out at a sweep rate of 0.16 mVs−1 from an initial potential of −0.4 V, with respect to the reference electrode, to 3.0 V when the scan was reversed.

The morphology of anodic layer has been evaluated by means of a Field Emission gun-scanning electron microscopy, FEG-SEM, and the qualitative analysis of the surfaces was performed by Energy Dispersive X-ray Spectroscopy (EDS).

2.2 Bacteria.

In order to carry out the microbiology experiment we have made use of two collection strains and six clinical strains. These collection strains were *S. aureus* 15981 [6] and *S. epidermidis* ATCC 35984. The collection strains are transformed cell lines, which are adapted to culture and maintain in the laboratory for long periods [7]. The clinical strains for *S. epidermidis* were P6, P33, P53B, P55, P74 and P101, while for *S. aureus* were P1, P2, P4, P18, P61T3 and P95. The clinical strains are bacteria that have been isolated by a sonication procedure [8] in the microbiology lab of the hospital Fundación Jiménez Diaz from patients who have had a biomaterial infection.

Journal of Physics: Conference Series **252** (2010) 012011 doi:10.1088/1742-6596/252/1/012011

2.3 Experiments
The staphylococcal adhesion experiments to the collection and clinical strains to modified and unmodified discs of Ti-6Al-4V was performed following the protocol developed by Kinnari et al [9]. We have employed *S. aureus* and *S. epidermidis* collection and clinical strains for these experiments. The experiments begins with a bacterial culture that was grown for 24 hours in TSB (Tryptic Soy Agar). After culture, bacteria is harvested by 10 min at 3500 rpm at 22 ° C. Supernatant is discarded and the pellet is washed three times with sterile PBS (Phosphate buffered saline). Bacteria is suspended and diluted in PBS with a 10^8 CFU/ml bacterial inoculum, which is incubated with discs of different modified surfaces of anodized titanium alloy (Ti-6Al-4V) during 90 minutes at +37ºC to allow adhesion. After incubation, discs are washed three times with PBS to remove unattached bacteria. Finally, discs are stained during 2 min with acridine orange and rinsed with water. Eight photographs (40 X magnification) in a fluorescence microscope for each disc are made. The surface area covered with adhered bacteria is calculated and studied by using ImageJ software (National Institute of Health Bethesda, MD), which allowed to calculate the mean percentage of biomaterial surface covered by each bacteria. The statistical analysis is performed using EPI-INFO software. The experiments are performed in triplicate for each collection strain.

2.3.1 *First experiment.* The adhesion to the collection strains to modified and unmodified discs of Ti-6Al-4V was performed following the above described protocol. We have used *S .aureus* 15981 and *S. epidermidis* ATCC 35984 for this experiment.

2.3.2 *Second experiment.* The adhesion of six clinical strains for *S. aureus* : P1, P2, P4, P18, P61T3 and P95 and six for *S. epidermidis* : P6, P33, P53B, P55, P74 and P101 to modified and unmodified discs of Ti-6Al-4V were performed following the same protocol. The experiments were performed in triplicate for each clinical strain.

2.3.3.*Third experiment*. The adhesion of collection strains of *S .aureus* and *S. epidermidis* to discs of Ti-6Al-4V anodized with in H_2SO_4 /HF and discs of Ti-6Al-4V anodized with H_2SO_4 acid was performed following the described protocol.

2.4. Statistical analysis.
The statistical study was performed using EPI-INFO software version 3.5.1 (CDC, EE.UU.). The results obtained were presented comparing the mean percentage of covered biomaterial surface for discs of Ti-6Al-4V, discs of Ti-6Al-4V anodized with H_2SO_4 /HF for 5 min and discs of Ti-6Al-4V anodized with H_2SO_4 /HF for 60 min. In order to carry out the statistical study, non-parametric tests were performed. When more than two samples were employed, the Kruskal-Wallis test was carried out, whereas for two samples, the Mann-Whitney test was used.

3. Results

3.1 Results of materials.
The morphology of the anodic oxide layer grown in the H_2SO_4 /HF anodising electrolyte varies depending on the anodising time. For the specimens treated for 5 minutes, the surface shows a nanotubular oxide film of 20 nm in diameter, approximately, while for longer anodizing times, 60 minutes, the nanotubes show about 100 nm in diameter and approximately 30 nm inter-spacing Figure1-2. The qualitative compositional analysis performed by EDS revealed fluorine content of 4 at % and 6 at % for the specimens treated for 5 and 60 minutes, respectively.

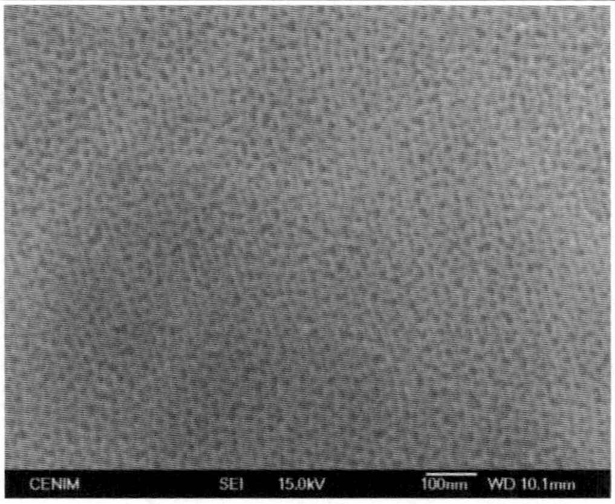

Figure 1: Morphology of the anodic oxide films grown in H_2SO_4 /HF anodizing bath. Nanotubular oxide films grown at 5 minutes

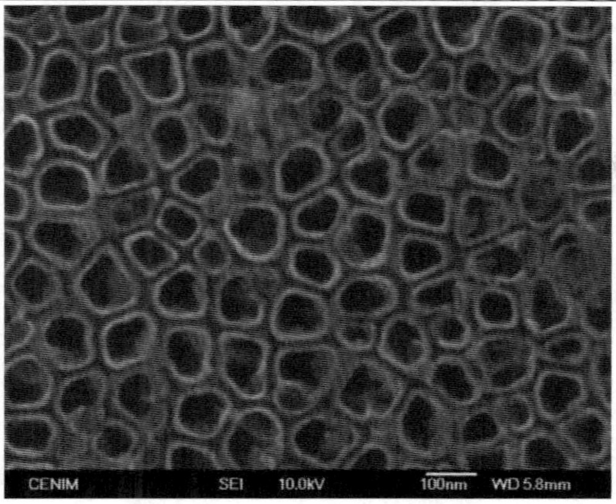

Figure 2: Morphology of the anodic oxide films grown in H_2SO_4 /HF anodizing bath. Nanotubular oxide films grown at 60 minutes.

The stability of the treated and untreated titanium surfaces were evaluated by potentiodynamic curves in PBS solution at 37ºC.The polarisation curves of chemically polished Ti-6Al-4V revealed a typical passivation behaviour with a corrosion potential of the forward scan of -0.414 V, passive current of 0.38 μA cm^{-2} and breakdown potential of the passive film of ~1.8 V (Fig. 2). Similar values have been describe in literature, 1.9 V, for mechanically polished Ti-6Al-4V in Hank's solution at 37ºC [10]. This breakdown potential is followed by steep current increase, associated with the pitting development. Subsequently, the reverse scan occurs at lower current densities, ~7.6 nA cm^{-2}. Very similar response was obtained for nanotube film formed for 60 min with slightly reduced hysteresis loop and positive shift of the corrosion potential by 0.11-0.12 V. The breakdown potential and passive current of the forward scan were ~1.87 V and 0.13 μA cm^{-2}, respectively. The passive current of the reverse scan was 1.2 nA cm^{-2}. Such similarity of the electrochemical responses indicates that electrolyte might penetrate through the inter-space of the nanotube walls reached the titanium substrate. The 5 min treatment nanotube film revealed the most stable passive behaviour, with a passive current density of 2 nA cm^{-2} and corrosion potential of 0.41 V; up to 3.25 V no pitting formation was observed.

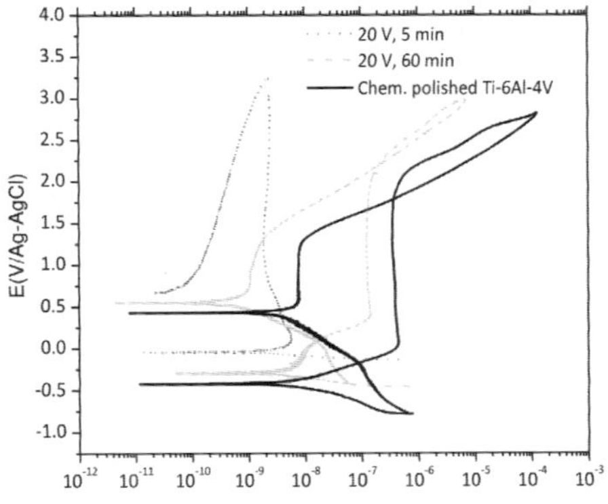

Figure 2: Cyclic polarisation curves for Ti-6Al-4V with and without nanotubular film in phosphate buffer solution at 37° C.

3.2 Results of adhesion of collection strains

The figure 3 shows the mean percentage of covered surface by *S. aureus* 15981 and *S. epidermidis* ATCC 35984 to discs of Ti-6Al-4V and discs of Ti-6Al-4V anodized with H_2SO_4 /HF for 5 min and discs of Ti-6Al-4V anodized with H_2SO_4 /HF for 60 min. These results show that the *S. aureus* adhesion is lower than *S. epidermidis* to modified and unmodified material.

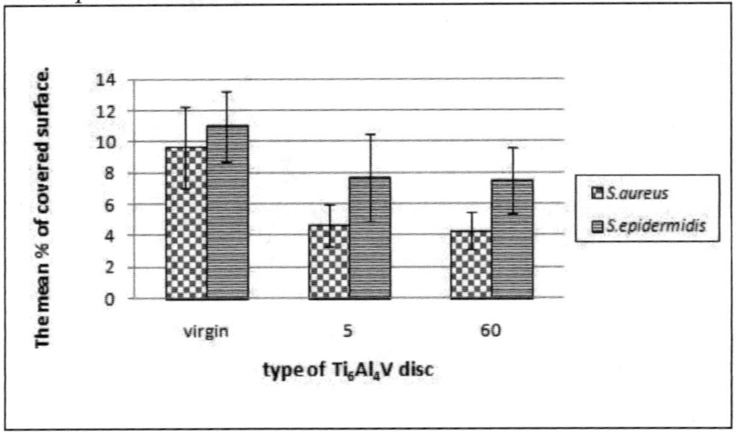

Figure 3. Results of collection strains adhesion to material surfaces; virgin (disc of Ti-6Al-4V), 5 (disc of Ti-6Al-4V anodized by H_2SO_4 /HF for 5 min) and 60 (disc of Ti-6Al-4V anodized by H_2SO_4 /HF for 60 min).

There are statistically significant differences between the behaviour to the two modified materials, for *S. aureus* the 5 and 60 minutes anodized disc, (both $p<0.0001$, Kruskal-Wallis test) and for *S. epidermidis* the 5min ($p= 0.0001$, Kruskal-Wallis test) and the 60 min anodized disc ($p<0.0001$, Kruskal-Wallis test) when compared to the unmodified material. There are no statistically significant differences among species when dealing with the disc of Ti-6Al-4V ($p<0.1293$, Kruskal-Wallis test). There are statically significant differences among species when testing the modified surface for 5 min ($p=0.0001$, Kruskal-Wallis test) and for 60 min ($p<0.0001$, Kruskal-Wallis test).

3.3 Results of adhesion of clinical strains.

The figures 4 and 5 show respectively the mean percentage of covered surface by six clinical strains of *S. aureus* and six clinical strains of *S. epidermidis* to disc of Ti-6Al-4V and discs of Ti-6Al-4V anodized with H_2SO_4 /HF for 5 min and discs of Ti-6Al-4V anodized with H_2SO_4 /HF for 60 min. We can see there is variability in the results among the strains of the same specie when comparing the covered surface for modified and unmodified discs of Ti-6Al-4V, both for *S. epidermidis* and *S. aureus.* There are differences in the adhesion between each strain of *S. aureus*. The P61T3 and P95 strains of *S .aureus* present a greater variability of covered surface when compared with the other strains. The P1 and P2 strains present a lower adhesion to modified material for 5 min when compared with the modified material for 60 min. The P1, P2 and P4 strains present a greater adhesion to modified material for 60 min than when compared with the P18, P61T3 and P95 strains Figure 4.

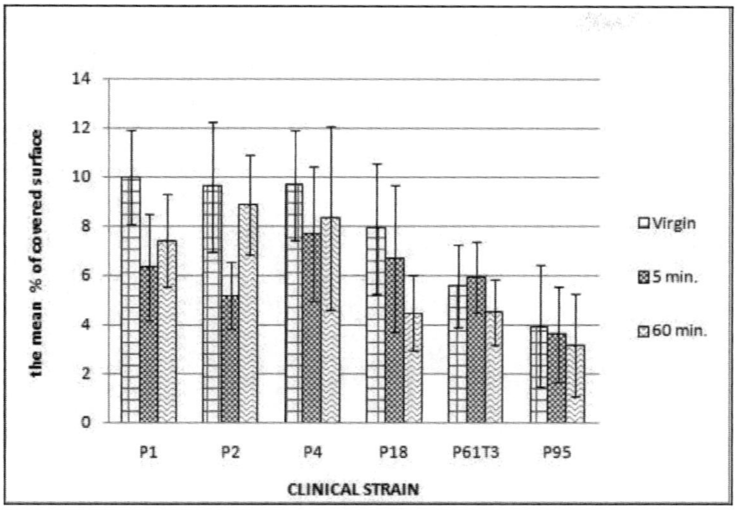

Figure 4.Results of *S. aureus* adhesion to material surfaces; virgin (disc of Ti-6Al-4V), 5 (disc of Ti-6Al-4V anodized by H_2SO_4 /HF for 5 min) and 60 (disc of Ti-6Al-4V anodized by H_2SO_4 /HF for 60 min).

There are statistically significant differences in the covered surface for the six clinical strains of *S. aureus* when comparing the modified material for 5 min and 60 min with the unmodified material ($p<0.0001$, Kruskal-Wallis test), however there are no statically significant differences when comparing the modified material for 5 min with the modified material for 60 min. *S. epidermidis* presents a smaller difference between the adhesion to modified and unmodified material compared with *S. aureus*. The main difference in the adhesion of *S. epidermidis* is found for the modified material for 5 min, while the modified material for 60 min presents, for P101 and P33 strains of *S. epidermidis*, a greater adhesion than for that of modified material for 5 min.

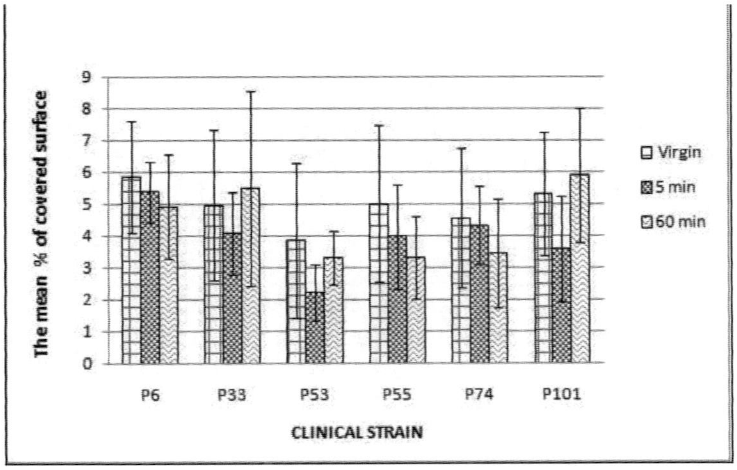

Figure 5.Results of *S. epidermidis* adhesion to material surfaces; virgin (disc of Ti-6Al-4V), 5 (disc of Ti-6Al-4V anodized by H_2SO_4 /HF for 5 min) and 60 (disc of Ti-6Al-4V anodized by H_2SO_4 /HF for 60 min).

There are statically significant differences in the covered surface for the six clinical strains of *S. epidermidis* when comparing the modified material for 5 min with the unmodified one ($p=0.0005$, Kruskal-Wallis test), and the modified material for 60 min ($p=0.0352$, Kruskal-Wallis test), when compared with the unmodified material. But no statically significant differences are found when comparing between the modified material for 5 min and 60 min (p=0.1742, Kruskal-Wallis test).

3.4 Results of adhesion of collection strains for discs with and without fluoride.
The figures 6 and 7 show the results of adhesion of collection strains to the modified material of Ti-6Al-4V by anodizing process with H_2SO_4 /HF and modified material of Ti-6Al-4V by anodizing process with H_2SO_4. There are statistical significant differences in the adhesion of *S. aureus* ($p< 0.001$, Kruskal-Wallis test) and *S. epidermidis* ($p= 0.0469$, Kruskal-Wallis test) if compared the adhesion to both materials. The adhesion of *S. aureus* diminishes around 50 % when comparing the covered surface between discs with fluoride and discs without fluoride Figure 6.

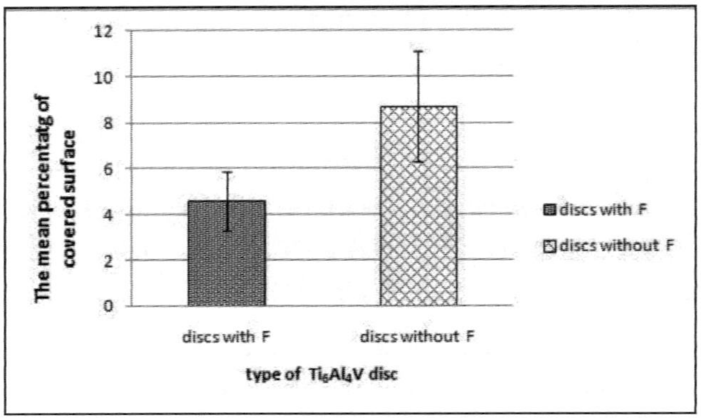

Figure 6. Results of *S. aureus* adhesion to disc of Ti-6Al-4V with fluoride and disc of Ti-6Al-4V without fluoride.

However, *S. epidermidis* presents less diminishment in the adhesion to discs with and without fluoride Figure 7.

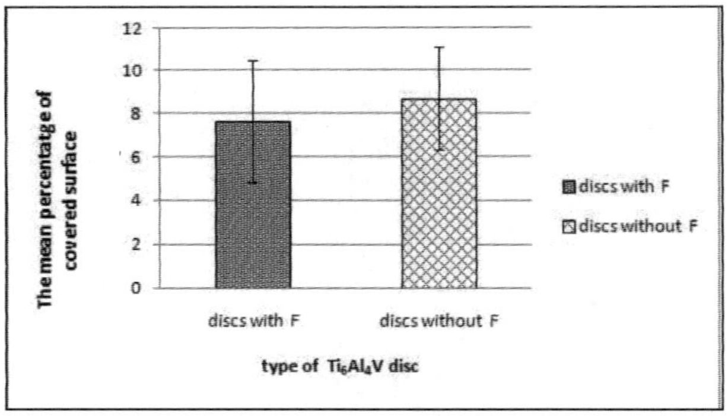

Figure 7. Results of *S. epidermidis* adhesion to discs of Ti-6Al-4V with and without fluoride.

4.Conclusions

Anodizing of Ti-6Al-4V alloy in sulphuric/hydrofluoric acid electrolyte, at controlled voltage, proceeds with formation of nanotubular oxide films. Anodizing for 5 min at 20 V results in formation of nanotubes of 15-20 nm in diameter. Increase of the anodizing time leads to formation of self-organised nanotubes of 100 nm in diameter. Nanotube film formed for 5 min demonstrated superior electrochemical stability with a passive current reduced by three orders of magnitude without an evidence of oxide breakdown up to 3.25 V compared with the substrate and 60 min film. Such behaviour is ascribed to a thicker continuous barrier layer and smaller pore size. Anodizing of Ti-6Al-

4V alloy in sulphuric/hydrofluoric acid electrolyte, at controlled voltage, proceeds with formation of nanotubular oxide films. Anodizing for 5 min at 20 V results in formation of nanotubes of 15-20 nm in diameter. Increase of the anodizing time leads to formation of self-organised nanotubes of 100 nm in diameter. Nanotube film formed for 5 min demonstrated superior electrochemical stability with a passive current reduced by three orders of magnitude without an evidence of oxide breakdown up to 3.25 V compared with the substrate and 60 min film. Such behaviour is ascribed to a thicker continuous barrier layer and smaller pore size.

The adhesion of bacteria to modified surfaces was decreased for the collection and clinical strains of both species. *S. epidermidis* had greater adhesion to both modified and unmodified material than *S. aureus*. There are statistically significant differences in both species adherence to modified and unmodified materials. The results also reflect a difference between both species in the adherence to the 5 min and 60 min modified material. The discs of Ti-6Al-4V following the anodizing process with H_2SO_4 /HF for 5 min show a lower adhesion for all the strains of both species. The results obtained could have a clinical application in decreasing the infection risk of patients receiving an implant.

5. Acknowledgements
The work was funded by a grant from the MCYT (CONSOLIDER-INGENIO 2010 CSD 2008-0023 FUNCOAT).

6. References

[1] Del Pozo J et al 2009 Infection Asociated with Prosthetic Joints. *N Engl J Med* **361**787 794
[2] Sherk H 1987 Biomaterial-centered infection *Clin Orthop Rel Res* **237** 1588 1595
[3] Katsikogianni M et al 2004 Concise Review of Mechanism of Bacterial Adhesion to Biomaterials and Techniques Used In Estimating Bacteria-Material Interactions. *Eur Cell Mater* **8** 37 57
[4] Yuehuei H et al 1998 Concise Rewiew of Mechanism of Bacterial Adhesion to Biomaterial Surfaces *J Biomed Mat Res* **43** 338 348
[5] Arenas MA, Conde A, Damborenea JJ, Matykina E, Esteban Moreno J, Gómez Barrena E, Perez- Jorge Permarch C, Pérez Tanoira R. MATERIALES DE TITANIO ANODIZADOS CON FLUOR. Patent: P201030720. Spain. CSIC- IIS– FUNDACION JIMENEZ DIAZ
[6] Valle J, Toledo-Arana A, Berasain C et al. 2003. SarA and not sigmaB is essential for biofilm development by *Staphylococcus aureus*. *Mol Microbiol;* **48** 1075 1087
[7] Kirkpatrick CJ, Mittermayer C. 1990. Theorical and Practical Aspects of Testing Potential Biomaterials in vitro.*J Mater Sci:Mater Med* **1** 9 13
[8] Esteban J et al 2008 Evaluation of Quantitative Analysis of Cultures from Sonicated Retrieved Orthopaedic Implants in Diagnosis of Orthopaedic Infection *J Clin Microbiol* **46** 488 49
[9] Kinnari TJ et al 2008 Adhesion of Staphylococcal and Caco-2 cells on diamond-like carbon polymer hybrid coating *J Biomed Mat Res Part A*. **86** 760 8.
[10] Gurappa, I. 2002 Characterization of different materials for corrosion resistance under simulated body fluid conditions. *Materials Characterization* **49** 1, pp. 73-79

Antibacterial efficacy of advanced silver-amorphous carbon coatings deposited using the pulsed dual cathodic arc technique

J. L. Endrino[1,*], A. Anders[2], J.M. Albella[1], J. A Horton[3], T. H. Horton[3], P. R. Ayyalasomayajula[4], M. Allen[5]

1 Instituto de Ciencias de Materiales de Madrid, Madrid, Spain
2 Lawrence Berkeley National Laboratory, Berkeley, California
3 SUNY Upstate Medical University, Syracuse, New York
4 Nano-Ram Technologies, Bangalore, India
5 Ohio State University, Columbus, Ohio

*E-mail: jlendrino@icmm.csic.es

Abstract. Amorphous carbon (a-C) also referred as diamond-like carbon (DLC) films are well known to be a biocompatible material with good chemical inertness; this makes it a strong candidate to be used as a matrix that embeds metallic elements with an antimicrobial effect. We have deposited a set of a-C:Ag films using a dual-cathode pulsed filtered cathodic arc source, the arc pulse frequency of the silver and graphite cathodes was controlled in order to obtain samples with various silver contents. In this study, we show the deposition of silver and carbon ions using this technique and analyze the advantages of incorporating silver into a-C by studying the antimicrobial properties against staphylococcus of samples deposited on Ti-6Al-4V coupons and evaluated using 24-well tissue culture plates.

Keywords. antibacterial, silver, filtered cathodic arc deposition, DLC, diamond-like carbon.

1. Introduction

Many orthopedic operations involve potentially contaminated surgical sites. Unfortunately, the use of orthopedic implants is associated with a relatively high incidence of infection. In addition to causing local pain, infections around surgical implants cause bone destruction that, ultimately, produce loosening of the implant and prevents the bone from healing. In an attempt to address this, advanced antibacterial coatings can be developed to prevent infections by decreasing the bacterial adhesion to the implant and that can also enhance bone growth around the implanted devices. DLC refers to an amorphous form of carbon in which the bonding is composed of a mixture of diamond type (sp^3) and graphite type (sp^2) hybridization. DLC has many properties in common with diamond, including high hardness, Young's modulus and chemical inertness. The material is of particular interest with regard to medical implants because it is highly biocompatible and can be deposited by physical vapor deposition (PVD) processes, which are done at much lower temperatures than chemical vapor deposited (CVD) diamond films. However carbon-based coatings do not possess any inherent antibacterial activity.

Metallic coatings such as silver can be effective in killing bacteria, but they can also produce a chronic inflammatory reaction that can lead to adverse tissue responses around a silver-coated device.

Table 1. Summary of some previous works in the literature related to antibacterial properties of modified DLC films.

Authors	Material	Deposition Process	Main Results
Endrino et al. [1]	a-C:Ag, a-C:H-Ag	Plasma immersion ion implantation and deposition (PIII&D)	No adverse cell growth of osteoblast MC3T3 cells in both hydrogenated and non-hydrogenated DLC containing 5 at. % Ag but non-hydrogenated samples had higher cell viability.
Hauert et al. [2, 3]	Ag/a-C:H	PACVD and magnetron sputtering	8 at.% Ag did not cause any change in cell growth rate of osteoblast MC3T3 cells but 16 at.% Ag decreased growth rate by a factor of two.
Ishihara et al. [4]	F-DLC	Reactive RF magnetron sputtering in mixed Ar, CH_4, CF_4 gas	Fluorine incorporation decreased the adhesion of *Escherichia Coli* bacteria to the DLC.
Katsikogianni et al. [5, 6]	Ag-DLC:F	Atom beam/magnetron sputtering.	The adhesion of *S. epidermidis* bacteria to the Ag-DLC coating was greater than that for Ag but less than that observed for DLC coating.
Kwok et al. [7]	Ag-DLC	Pulsed filtered cathodic vacuum Arc	Samples deposited with very high Ag concentration on the surface (60-80 at.%). DLC-Ag films proved effective against *Escherichia coli* bacteria.
M.L Morrison et al. [8]	Ag-DLC Pt-DLC AgPt-DLC	Pulsed laser deposition (PLD)	Doping of diamond-like carbon with silver and platinum had increased antimicrobial efficacy against *Staphylococcus warneri.*
Narayan et al. [9]	Ag-Pt-DLC	Pulsed laser deposition (PLD)	Antimicrobial efficacy of DLC-Ag-Pt reported against Staphylococcus and *Pseudonomas aeruginosa* bacteria.

Thus, silver-containing diamond-like carbons have attracted a lot of attention in the field of antibacterial coatings due also to promising antibacterial (see Table 1) and tribological properties [10]. Generally speaking, silver-containing DLC materials can be divided in two groups: hydrogenated and non-hydrogenated DLC, which is dependent on the deposition technique. In our recent work [1], we carried out a comparative set of depositions using either pulsed dual-cathode filtered cathodic vacuum arc (PDC-FCVA) to obtain non-hydrogenated samples and plasma based ion immersion implantation and deposition (PBIIID) using methane gas as precursor to obtain a hydrogenated matrix (see videos in Fig. 1). Our result concluded that the non-hydrogenated DLC matrix was better because they promoted higher cell viability of mouse osteoblastic cells and therefore from a biomedical point of view are a preferred choice for the preparation of antibacterial coatings.

Journal of Physics: Conference Series **252** (2010) 012012 doi:10.1088/1742-6596/252/1/012012

2. Experimental

2.1 Deposition method

Amorphous carbon films doped with silver were prepared on glass disc substrates by PDC-FCVA. The deposition system consisted of a dual-cathode "triggerless" mini-gun designed to operate in pulsed mode [11]. Pulsed arc discharges on individual cathodes were triggered by a computer control system. The pulses used in the production of carbon plasma were 100 A and had a duration of 5 ms, while the metal plasma was made using pulses 1 ms long and a current of 700 A. The plasma stream produced by the source is injected into a 90-degree filter to remove most of the macroparticles which were formed during the cathodic arc process, the output of the filter can be observed in the videos of figures 1a and 1b. After exiting the filter, the plasma expanded through a homogenizer device composed of concentric magnets and allowed to have homogeneous thickness and composition in the substrates over an area of 80 cm^2. During the deposition of carbon pulses, the substrates were biased with negative 1 kV pulses that were 2 μs long with an off time between pulsed of 14 μs. The substrate was kept at ground potential during the deposition of the metal plasma (hence "species selective bias") [12]. The pulse repetition rate was set at 2 pulses per second and the number of pulses per deposition was 5000. During the deposition, the substrate holder was rotated at a speed of 2 revolutions per minute (RPM). The residual gas pressure was typically in the 10^{-4} Pa range. The thicknesses of the deposited films ranged from 50 nm to 300 nm.

(a) (b)

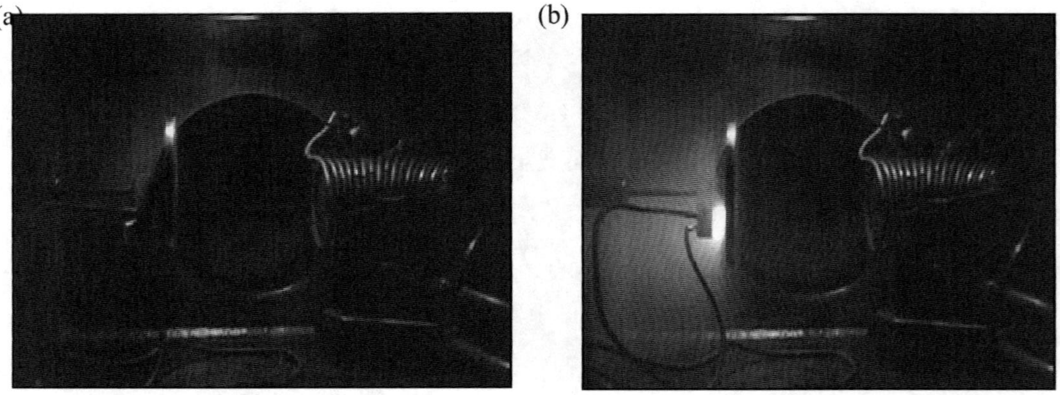

Figure 1.Videos of the deposition processes used to obtain non-hydrogentated and hydrogenated DLC-Ag films. Plasma homogenizer between filter output and rotating substrates was removed for illustrating purposes. (a) Video of pulsed filtered cathodic arc alternating graphite and silver pulses. (b) Video of plasma based ion implantation and deposition of methane simultaneously with pulsed filtered cathodic deposition of silver (process shown here only for comparative purposes). The violet plasma discharge around and behind the substrate holder indicates the continuous implantation from methane gas. Note: videos are already embedded in the pdf document. Adobe reader with adobe flash player is needed to watch the videos. Security settings should allow the embedded functionality.

2.2 Surface energy

A KRUSS DAS 100 instrument was used for contact angle measurements. This computer-controlled contact angle analyzer allowed us to measure static and dynamic contact angle along with surface tension and surface energy. The position of the sample stage can be precisely adjusted along the x-, y-, or z-axis. An optical system using variable intensity illumination controls the degree of backlighting. The optics can be tilted by up to 3° allowing us to choose the angle of view required for rough or uneven surfaces. The instrument contains a stepping motor controlled syringe system for precise and repeatable liquid drop formation and application. This insures that a reproducible drop volume is applied to the surface. The micro-syringe attachment facilitates rapid and precise formation of drop specimens. Following placement of a sessile drop, the syringe is easily swung from the field of view.

It will return to its exact working location for placement of the subsequent drop. Alternatively, it can remain at the working location, allowing continuously adding or withdrawing fluid from a sessile drop for the purpose of forming advancing or receding contact angles. After placing the water droplet, the contact angle was measured using the circular curve fit option of the software.

2.3 Antibacterial tests
Sterilized glass discs with the indicated a-C:Ag surface composition were transferred to 12-well polystyrene culture plates, and pre-treated in 1 ml DMEM media (DMEM is the abbreviation for Dulbecco/Vogt modified Eagle's (Harry Eagle) minimal essential medium) supplemented with 10% fetal calf serum (CellGro, Herndon, VA, USA) 24 hours at 37°C with agitation at 100 RPM. Following pre- treatment, 100 μl aliquots of the media were reserved for subsequent verification of initial sample sterility. No pre-existent microbial contamination was observed after 48 hours of incubation onto solid BHI agar.

Lyophilized cultures of *Staphylococcus aureus* (ATCC #25923, Manassas VA, USA) were revived in Brain-heart Infusion liquid media (BBL-BHI broth, Becton, Dickinson and Co., Sparks, MD, USA) for 6 hours at 37°C with agitation at 220 RPM. From this culture, 100 μl were applied to Brain Heart Infusion solid agar media (BBL-BHI agar, Becton, Dickinson and Co.) and incubated for 36 hours at 37°C. For each experiment, a single colony (~1.5 mm) was selected from the agar plate and dispersed in 12.5 ml Dulbecco's modification of Eagle's media supplemented with 10% fetal calf serum (DMEM/10%FCS). The DMEM/10% FCS cell suspension was allowed to grow 3-4 hours at 37°C with agitation at 220 RPM until a spectrophotometric optical density of approximately 0.5 AU was achieved at 670 nm wavelength. Subsequently, the suspension volume was adjusted with fresh media to normalize the inocula for each replicate experiment to 0.3 AU at 670 nm, yielding on approximately 1.64 $x10^7$ CFU/ml after incubating a 16-fold diluted suspension on BHI agar for 36 hours at 37°C.

The remaining media was aspirated following pre-treatment of the a-C:Ag discs, replaced with 1 ml of the DMEM/10% FCS cell suspension and incubated for 24 hours at 37°C with agitation at 100 RPM. After incubation, 500 μl of media from each well was collected, transferred to a sterile polypropylene microcentrifuge tube and stored briefly on ice. To quantify bacterial viability on solid media, 1 μl of the collected cell suspension was diluted into 1 ml sterile PBS, and from this dilution, 10 μl was spread over the surface of a BHI agar plate. After 24 hours of growth at 37°C, a colony number was estimated, and from the number of colonies used to express the number of viable cells number in colony forming units/μl (CFU). The remaining media was aspirated, and the discs were washed three times in sterile phosphate buffered saline (PBS, pH 7.4) to remove cells that were not adhered to the surface of the DLC:Ag discs. Following the final wash, the discs were stained for 15 minutes at room temperature in a solution of 10 μM Sytox-9 and 60 μM propidum iodide (BacLight Live/Dead Kit, Invitrogen, Carlsbad, CA, USA) dissolved in Phosphate buffered saline (PBS) as recommended in the manufacturer's protcol. Subsequently, the discs were transferred to glass microscope slides and coverslipped with the supplied mounting oil. Digital micrographs were captured through a 100x oil-immersion objective under epifluorescent illumination using FITC (Ex.480nm/Em.535) or G2A (Ex.510nm/Em590nm) dichroic filter sets to respectively visualize Sytox-9 (green) or propidum iodide (red) stained cells. Corresponding monochromatic images were processed and merged using Image Pro Plus software (V4.0, Media Cybernetics, Silver Spring, MD, USA).

3. Results

3.1 Wettability
The ratio of silver and carbon pulses was varied in order to obtain samples with various silver contents. Figure 2 shows the contact angle of all the deposited samples. The specimens deposited using ratios of 1 pulse of silver for 12 pulses of carbon or less had overall higher surface energies (i.e. lower contact angles). The lowest contact angle was obtained by the film deposited using 1 pulse of

silver for 12 of carbon. Samples containing higher silver contents had a hydrophilic/hydrophobic behaviour with contact angles slightly below 100 degrees. These samples had very similar surface energies which did not change even in the case of the pure silver coating, indicating the high presence of the metal on the surface of samples deposited with 6:1 and 3:1 pulse ratios.

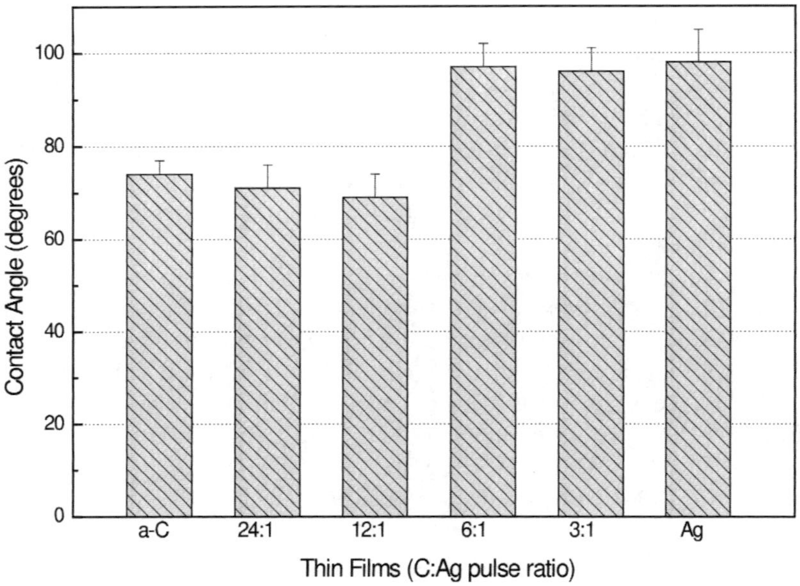

Figure 2. Contact Angle of various a-C:Ag films

3.2 Surface biocidal effect

As reported in the experimental section, a BacLight live/dead assay was used in order to determine the biocidal effect on the surface of the films prepared with increasing contents of silver. The principle of this assay relies upon the differential staining properties of live bacteria versus dead bacteria. All cells were observed to stain with the cell permeable Sytox-9 fluorochrome dye, while cells simultaneously staining with the cell-impermeable DNA intercalator propidium iodide fluorescence were interpreted to be dead (tainted in red). As can be observed in the microscope fluorescence images of figure 3, there appeared to be a dose dependent decrease in both the total number of adherent cells as well as the percentage of viable cells adherent to the a-C:Ag coated discs. The number of dead bacteria on the surface is higher than the number of live bacteria in figures 3(d)-(f). These were also the samples with higher contact angles and higher concentration of silver.

3.3 Antibacterial efficacy on the medium

Microorganisms can colonize the surfaces of medical devices and form matrixes of bacteria and proteins called biofilms. In order to determine the possible biofilm inhibition effect of the prepared a-C:Ag films on the medium, a fluorometric assay was used to quantify the percent viability of bacteria in suspended medium (Fig. 4). There was an inverse association (r^2=0.980) between bacterial viability in suspension and the ratio of silver pulses used to coat the discs, however, statistically significant reductions in the percent viability with respect to the a-C discs of cells in suspension were evident at the highest concentration of Ag ($p<0.0001$). Similarly, when the suspended medium was transferred to solid media for colony counting, cultures derived from the pure Ag sample (0.28±0.39 CFU/μl) showed a significant reduction in colony forming units relative to the pure a-C coating (43.14±11.78 CFU/μl, $p<0.0001$), as shown in Fig. 5. However, there was no statistically significant reduction of colony forming units in mixed a-C:Ag films in comparison with pure a-C film.

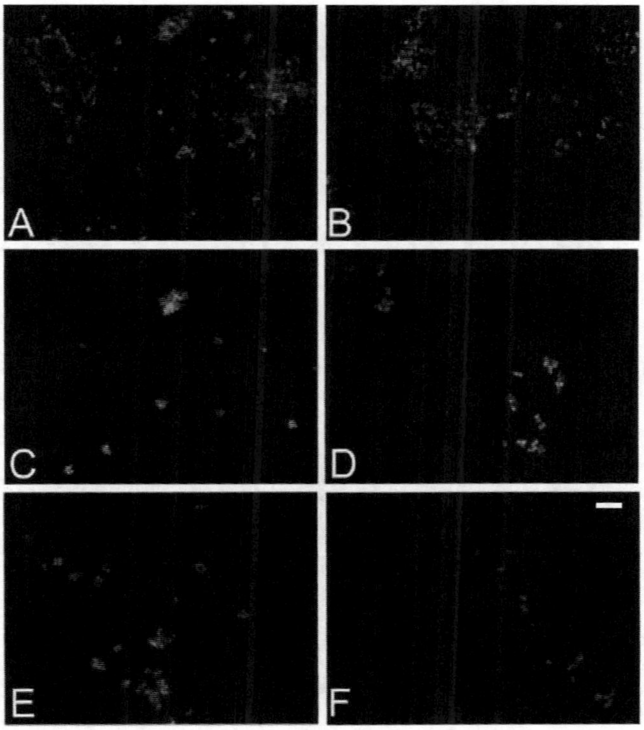

Figure 3. Fluorescent microscopy of *Staphylococus.aureus* adherent to a-C:Ag coated TiAlV discs at ratios of (a) a-C (b) 24:1, (c) 12:1, (d) 6:1, (e) 3:1, and (f) pure Ag. Monochromatic digital micrographs were individually captured with a 100x oil-immersion objective under epifluorescent illumination at either Ex.480nm/Em.535nm (green) to visualize viable cells, or Ex.510nm /Em590nm (red) to visualize dead cells and corresponding images were overlayed. Scale bar = 10μm.

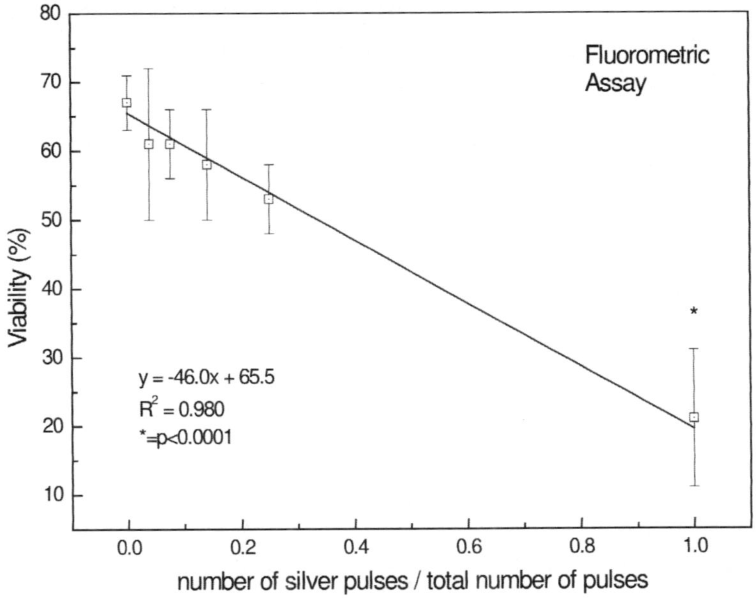

Figure 4. Fluorometric assay of suspended *S. aureus* viability after 24 hours of culture versus a-C:Ag coated discs with varying Ag pulse ratios. Mean of n=4 replicate experiments; bars show 1 standard deviation (SD).

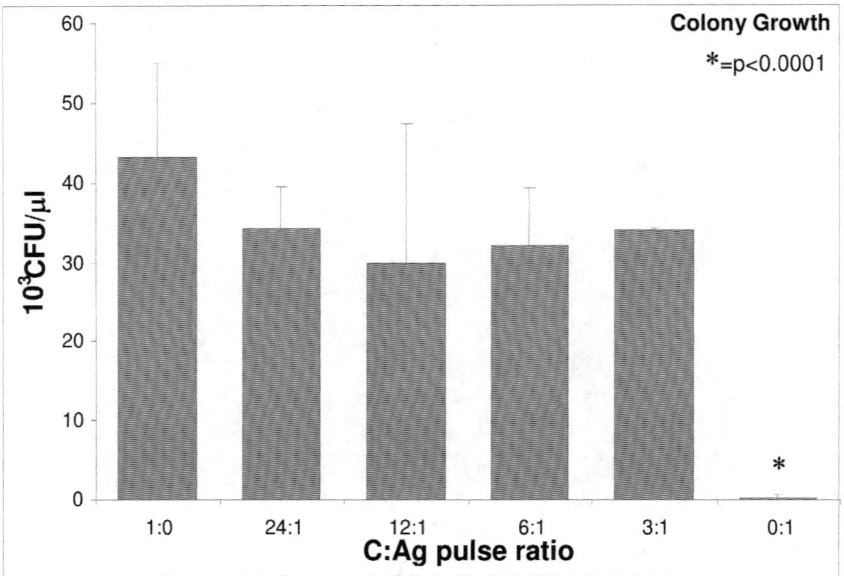

Figure 5. *Staphylococcus aureus* colony forming units in suspension after 24 hours of culture with a-C:Ag coated discs with varying pulse ratios. Mean of n=2 replicate experiments; bars show 1 standard deviation (SD).

4. Conclusions and future work

In this study, the antimicrobial property of hydrogen-free a-C:Ag films, deposited by the pulsed dual-cathode filtered cathodic arc technique, was analyzed in both the surface of the prepared thin films and in the medium. Microscopically, there was a clear correlation between the number of viable Staphylococcus bacteria adhered to the surface and the decreased wettability of the films due to the increased presence of silver in the surface. However, even when there was a drop in the amount of viable cells in the medium in some mixed a-C:Ag films, there was only statistically significant reduction in the number of colony forming units in the medium for the case of the pure silver film. In view of these reported antibacterial results as well as the results reported in other studies on DLC-Ag films, future efforts on the development of silver containing amorphous carbons should focus on studies of biofilm formation as well as in-vivo tests.

5. Acknowledgements

The authors thank J. C. Sánchez-López (ICMSE, CSIC) for his helpful comments. This work was partially supported by MICINN project Consolider (CSD 2008-0023 FUNCOAT). AA acknowledges support by the U.S. Department of Energy under Contract No. DE-AC02-05CH11231.

6. References

[1] Endrino J L, Escobar Galindo R, Zhang H S, Allen M, Gago R, Espinosa A and Anders A 2008 Structure and properties of silver-containing a-C(H) films deposited by plasma immersion ion implantation *Surface and Coatings Technology* **202** 3675-82

[2] Hauert R 2003 A review of modified DLC coatings for biological applications *Diamond and Related Materials* **12** 583-9

[3] Hauert R, Gampp R, Muller U, Schroeder A, Blum J, Mayer J, Birchler F and Wintermantel E 1997 Surface analysis and bioreactions on silver-containing amorphous hydrogenated carbon

films *American Chemical Society, Polymer Preprints, Division of Polymer Chemistry* **38** 994-5

[4] Ishihara M, Kosaka T, Nakamura T, Tsugawa K, Hasegawa M, Kokai F and Koga Y 2006 Antibacterial activity of fluorine incorporated DLC films *Diamond and Related Materials* **15** 1011-4

[5] Katsikogianni M, Spiliopoulou I, Dowling D and Missirlis Y F 2004 Bacterial (S. Epidermidis) adhesion on P VC and D LC/Ag/CF_4 coatings. I n: *Transactions - 7th World Biomaterials Congress,* (Sydney) p 1082

[6] Katsikogianni M G , S yndrevelis C S, Amanatides E K, Mataras D S and Missirlis Y F 2007 Plasma t reated and a-C:H coated P ET p erformance i n i nhibiting ba cterial a dhesion *Plasma Processes and Polymers* **4**

[7] Kwok S C H, Z hang W , W an G J, M cKenzie D R, B ilek M M M a nd C hu P K 2007 Hemocompatibility and anti-bacterial properties of silver doped diamond-like carbon prepared by pulsed filtered cathodic vacuum arc deposition *Diamond and Related Materials* **16** 1353-60

[8] Morrison M L, Buchanan R A, Liaw P K, Berry C J, Brigmon R L, Riester L, Abernathy H, Jin C and Narayan R J 2006 Electrochemical and antimicrobial properties of diamondlike carbon-metal composite films *Diamond and Related Materials* **15** 138-46

[9] Narayan R J, Abernathy H, Riester L, Berry C J and Brigmon R 2005 Antimicrobial properties of di amond-like car bon-silver-platinum na nocomposite t hin f ilms *Journal of Materials Engineering and Performance* **14** 435-40

[10] Zhang H S, Endrino J L and Anders A 2009 Modification of surface and tribological properties of DLC films by adding silver content. In: *2008 Proceedings of the STLE/ASME International Joint Tribology Conference, IJTC 2008,* pp 537-9

[11] Anders A, Brown IG, McGill RA, and Dickinson MR 1998 "Triggerless" triggering of vacuum arcs," J. Phys. D: Appl. Phys. **31**, 584-587

[12] Anders A., P asaja N., a nd S. S ansongsiri S Filtered cathodic a rc de position w ith i on-species-selective bias, Rev. Sci. Instrum. **78**, 063901-1-5

Journal of Physics: Conference Series **252** (2010) 012013 doi:10.1088/1742-6596/252/1/012013

Bacterial adherence on fluorinated carbon based coatings deposited on polyethylene surfaces

A Terriza[1], G del Prado[2], A Ortiz Pérez[2], M J Martínez[3], J A Puértolas[3], D Molina Manso[2], A R González-Elipe[1], F Yubero[1], E Gómez Barrena[2] and J Esteban[2]

[1]ICMSE (CSIC- Univ. Sevilla), Américo Vespucio 49, E-41092 Sevilla, Spain. http://sincaf-icmse.es
[2]IIS-Fundación Jiménez-Díaz, Madrid, Spain.
[3]ICMA (CSIC-Univ. Zaragoza) C/ María de Luna 3, E-50018 Zaragoza, Spain.

E-mail: antonia.terriza@icmse.csic.es

Abstract. Development of intrinsically antibacterial surfaces is of key importance in the context of prostheses used in orthopaedic surgery. In this work we present a thorough study of several plasma based coatings that may be used with this functionality: diamond like carbon (DLC), fluorine doped DLC (F-DLC) and a high fluorine content carbon-fluor polymer (CF_X). The study correlates the surface chemistry and hydrophobicity of the coating surfaces with their antibacterial performance. The coatings were deposited by RF-plasma assisted deposition at room temperature on ultra high molecular weight polyethylene (UHMWPE) samples. Fluorine content and relative amount of C-C and C-F bond types was monitored by X-ray photoelectron spectroscopy and hydrophobicity by water contact angle measurements. Adherence of *Staphylococcus aureus* and *Staphylococcus epidermidis* to non-coated and coated UHMWPE samples was evaluated. Comparisons of the adherence performance were evaluated using a paired *t* test (two materials) and a Kruskall Wallis test (all the materials). *S. aureus* was statistically significant ($p \leq 0.001$) less adherent to DLC and F -DLC surfaces than *S. epidermidis*. Both bacteria showed reduction of adherence on DLC/UHMWPE. For *S. aureus*, reduction of bacterial adherence on F-DLC/UHMWPE was statistically significant respect to all other materials.

1. Introduction.

Bacterial adherence is the process whereby bacteria attach themselves to cells or other surfaces before proliferating. Development of intrinsically antibacterial surfaces is of key importance in the context of prostheses used in orthopaedic surgery. Bacterial adherence on implant surfaces is the first step in the infection process that can eventually lead to implant removal. Septic failure of prostheses represents severe consequence not only because morbidity and mortality of patients, but also for the society, which faces high economical burdens.

The bacterial adherence process is a complex series of physical and chemical interactions between the substratum and the microbe, and thus can be modified with changes in the biomaterial under risk.

One approach to provide anti-adherent functionality is to manipulate the properties of biomaterials' surface by coating them with specific chemical functionalities.

Diamond like carbon (DLC) and other coatings (as, for example, F-DLC, Si-DLC, or Ag-DLC) have been proposed to improve the antibacterial performance of biomaterials [1]. It is also worth mentioning that DLC coatings have shown good performances regarding hardness, wear resistance, friction, chemical inertness and good biocompatibility, which make them a good option for implants in biomedical applications [2].

Attending to the previous hypothesis, we have 1) selected the microorganisms to be tested from two staphylococci species, *Staphylococcus aureus* and *Staphylococcus epidermidis*, both usual gram positive bacteria which are involved in the most of biomaterial-related infections and frequently isolated from orthopaedic infections; and 2) set the aims of the present study in the investigation of these two bacterial species adherence to carbon and fluorinated carbon based coatings on UHMWPE, compared to uncoated material.

2. Materials and methods.

2.1. UHMWPE.

The raw material used in the first part of the study was a compression moulded sheet of GUR 1050 UHMWPE (Orthoplastic Ltd., Lancashire UK). Slices of a size of 10x10 mm and ~600 microns thick were used as reference samples for the adhesion studies and as substrates for the coatings as described below. The roughness of the UHMWPE slices was in the level of the tenth of micron.

2.2. DLC, F-DLC, and CF_X coatings.

All the UHMWPE coatings were deposited at room temperature using a RF capacitive coupled plasma reactor. Details about the physical characteristics of the DLC deposited material can be found elsewhere [3]. Besides, two types of C-F functionalities were incorporated to the UHMWPE surface. A first one consisted of a DLC based material with a certain amount of F incorporated in its structure (F-DLC in the following). This was achieved by letting a certain amount of C_4F_8 gas in the reactor chamber while the DLC growth was taken place. A second one consisted of fluorine rich amorphous carbon structure obtained by simple decomposition of C_4F_8 precursor in a Ar plasma (CF_X in the following). Process time to coat the UHMWPE surface was about 10 minutes.

2.3. Surface chemical characterization.

Identification of the functional groups at the surface of the UHMWPE samples before and after surface processing was performed by X-ray Photoelectron Spectroscopy (XPS). The measurements were performed using a VG-ESCALAB100 electron spectrometer with unmonochromatised Mg Kα radiation. Pass energies of 20 eV or 100 eV were used for high resolution or survey spectra acquisition, respectively.

2.4. Contact angle measurements.

Contact angle (CA) measurements were performed in the last specimens with a CAM100 optical system (KSV Instruments, Helsinki, Finland) using deionised water.

2.5. Bacterial adherence study.

For bacterial adherence studies, two collection staphylococci strains were used: *S. aureus* 15981 [4] and *S. epidermidis* ATCC 35984. An inoculum of 10^7-10^8 colony former units (CFU) / mL was prepared according to a previously described method [5] as follows. Overnight-grown cultures in Tryptic-Soy Broth (bioMerieux Spain S.A., Madrid, Spain) at 37 ºC were centrifuged and the concentrates washed three times with Phosphate Saline Buffer (PBS, Sigma-Aldrich Inc., St. Louis, USA). Then, inoculum was prepared from concentrate, making a bacterial suspension in PBS of 0.5 Mc Farland turbidity. Inoculum size was confirmed with serial bacterial counts on Tryptic-Soy Agar supplemented with 5 % sheep blood (bioMerieux Spain S.A., Madrid, Spain).

Six well-tissue culture treated plates (Iwaki. Asashi Glass Co. LTD., Japan) were used as solid support for fixing the different accretion surfaces, which were previously sterilized by mean of gas plasma treatment. The inoculum was added in a final volume of 4 mL / well and plates were then incubated 90 minutes at 37 °C. After this incubation period, inoculum was removed and surfaces washed with PBS three times to elute non-adhered bacteria. Surfaces were incubated 24 hours at 4 °C.

Adhered bacteria were visualized by fluorescence microscopy (Laborlux D, Leitz, Germany) after Acridine Orange staining (AO, Becton Dickinson and Company, Sparks Maryland, USA). Surfaces were exposed to 4 mL of AO 2 minutes at room temperature. Excess stain was removed with distilled water washing for two times, and then, surfaces were air-dried. Microscopy visualization was performed at 40x magnification and images of 10-20 fields / surface were photographed.

The percentage of surface covered by bacteria was determined using *Image J* software [6]. Differences among surfaces as well as between strains were analysed by ANOVA, Kruskal-Wallis, Mann-Withney, unpaired and paired *t* tests. A statistical difference was considered for $p < 0.001$.

Assays were made in triplicates. Number of observations was of twelve in the case of UHMWPE surfaces and twenty for all of the other tested materials.

3. Results.

3.1. Chemical characteristics of the UHMWPE coated surfaces.

The chemical nature of the species present at the surface of the samples was analysed by XPS. Figure 1 shows the C 1s spectra acquired for the studied samples. Uncoated and DLC coated UHMWPE samples presented a single C 1s peak, that was located at 285.0 eV binding energy (BE). It corresponds to the aliphatic C bond type of polyethylene and also to the amorphous structure of sp^2/sp^3 bond types that compose the DLC material.

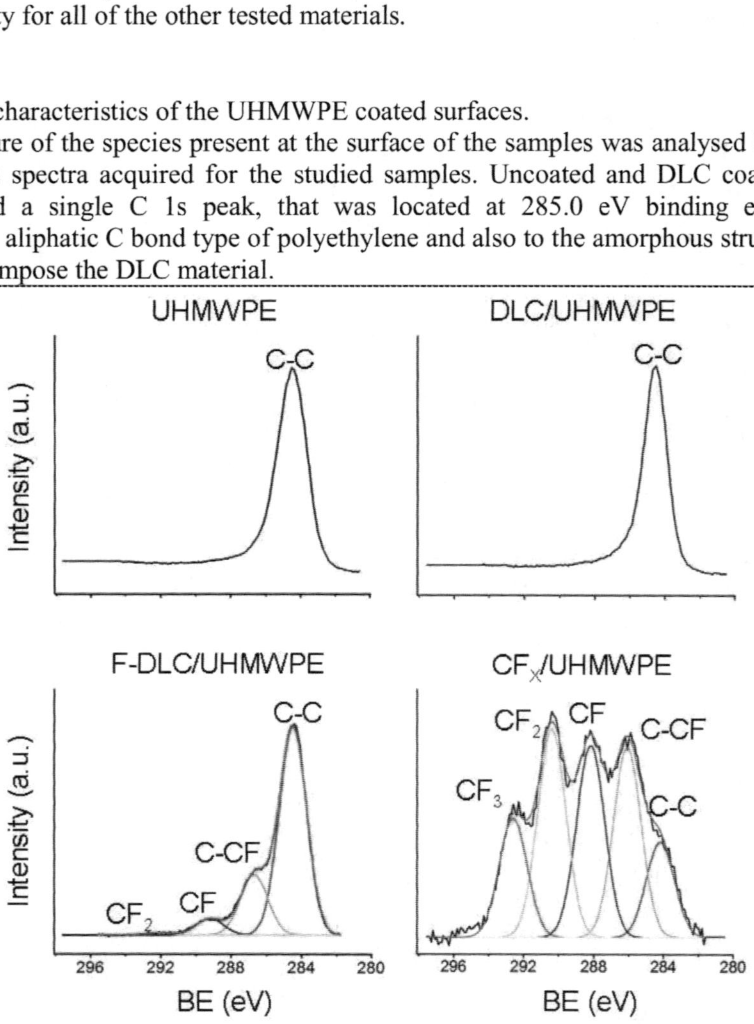

Figure 1. High resolution C 1s spectra of the samples obtained from the XPS analysis. The signal corresponding to the different C bond types present at the surface of the samples was deconvoluted by peak fitting analysis.

The addition of the fluorinated volatile precursor to the plasma process gas used for the DLC growth has the effect of incorporating some F content in the films (table 1) up to a [F]/[C] concentration ratio of 0.32. The F incorporated this way is mainly in the -C-CF form (21 %) , but also some minor contributions of -CF (5%) and -CF_2 (1%) is also detected.

The direct decomposition of the C_4F_8 precursor in a Ar plasma induces a completely different C-F based polymer. In this case, the F content increase considerably to a [F]/[C] ratio of 1.25, with the presence of all the possible -C-F bond types with a more or less even contribution of all the possible -C-F bond types (12% of -C-C-, 23 % of -C-CF, 24 % of -CF, 26 % of -CF_2, and 15 % of -CF_3).

Table 1. Quantification of C 1s peaks of UHMWPE surface before and coating with DLC, F-DLC and CF_X functionalities. Relative [F]/[C] atomic concentration of the samples is also included.

Sample	[F]/[C]	-C- bond types (%)				
		-C-C-	-C-CF	-CF	-CF_2	-CF_3
Raw UHMWPE	0	100	0	0	0	0
DLC/UHMWPE	0	100	0	0	0	0
F-DLC/UHMWPE	0.32	73	21	5	1	0
CF_x/UHMWPE	1.25	12	23	24	26	15

3.2. Surface wettability.

Water contact angle (WCA) measurements were performed to probe the hydrophobic character of the samples. The measurements were performed 24 hours after surface treatment to minimized the well-known aging effects after plasma surface treatment. The results of this analysis are compiled in table 2. The uncoated UHMWPE material presented a WCA of 100º, the DLC coated material 79º, the F-DLC coated 98º and the CF_X coated surface resulted to be superhydrophobic, i.e., with a WCA larger that 170º.

Table 2. Water contact angle (WCA) of virgin and coated UHMWPE samples. The standard deviations of the obtained mean are indicated with parenthesis.

Sample	WCA (degrees)
Raw UHMWPE	100 (5)
DLC/UHMWPE	79 (3)
F-DLC/UHMWPE	98 (4)
CF_x/UHMWPE	>170 (5)

3.3. Bacterial adherence studies.

Bacterial inocula were of 7.73 ± 0.18 and 7.84 ± 0.03 Log $_{10}$UFC/mL for *S. aureus* and *S. epidermidis*, respectively. After inoculum exposure, both bacteria were able to adhere to all of the accretion surfaces tested in the study. Figure 2 summarizes the percentage of bacterial-covered surface on the four materials. Between species, *S. epidermidis* was a much more adherent microorganism than *S. aureus*, because of the higher extent of covertures obtained in three out of the four tested surfaces. Statistically significant differences ($p \leq 0.001$) where found when compared percentages of covered surface by *S. aureus* and *S. epidermidis* in DLC/UHMWPE (2.18 ± 0.47 and 4.46 ± 1.03, respectively) and F-DLC/ UHMWPE (1.02 ± 0.27 and 7.00 ± 1.76, respectively).

In the comparison of bacterial adherence to each surface, the two strains showed a diminution in their adherence on DLC/UHMWPE surfaces. Statistically significant differences in *S. aureus* adherence were found in DLC/UHMWPE versus UHMWPE (2.18 ± 0.47 vs. 4.43 ± 1.50) and versus CF_x/UHMWPE (2.18 ± 0.47 vs.4.28 ± 1.47). Significant differences in *S. epidermidis* adherence were also found in UHMWPE versus DLC/UHMWPE (7.21 ± 2.95 vs. 4.46 ± 1.03) and CF_x/UHMWPE versus F-DLC/UHMWPE (4.05 ± 1.14 vs. 7.00 ± 1.76).

The lowest percentage of bacterial adherence was registered for F-DLC/ UHMWPE surfaces exposed to *S. aureus* inocula (1.02 ± 0.27). On the contrary, *S. epidermidis* adherence to F-DLC/UHMWPE was similar to that observed in UHMWPE.

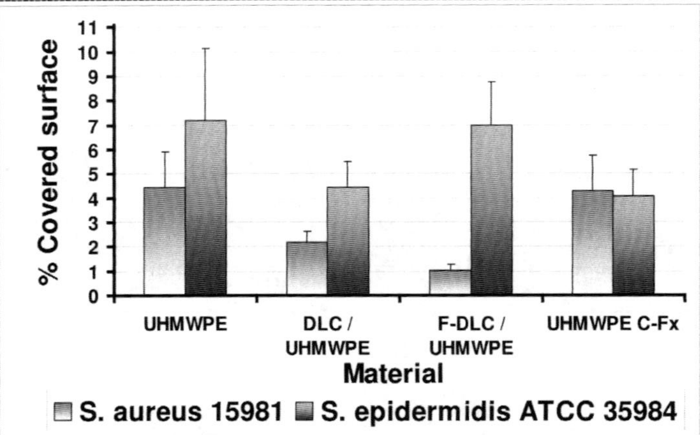

Figure 2. Percentage of covered surface by *S. aureus* 15981 and *S. epidermidis* ATCC 35984 on UHMWPE and modified UHMWPE surfaces (the error bars indicate the standard deviation of the corresponding mean values).

4. Discussion.

Bacterial adherence and subsequent biofilm formation are a spontaneous, natural phenomenon which depends on multiple factors including the microorganism, inoculum size, environmental physic-chemical conditions (atmosphere, pH, temperature), nutrient availability, hydrodynamic forces, or the nature of accretion surface [7]. Bacterial adherence is a complex phenomenon which involves two different phases where different relationships are established between the organism and the surface. Because no changes can be made on bacteria, modifications of surfaces have been suggested to be a interesting strategy for preventing bacterial adherence and biofilm development. *In vitro* studies involving the surface coating with biological molecules or compounds [8,9] and, further, *in vivo* experimental assays using coating medical devices [10,9], have been previously described with promising results. One of the explanations about the properties of these compounds for preventing bacterial adherence is based on the changes in accretion surface's charge and thus, the changes in interaction between bacteria and surface [11]. According to this premise, here we show that the inert DLC coatings are more hydrophilic that the uncoated polyethylene surfaces. It has been reported that nutrient concentration could be favored by apolar surfaces, and such phenomenon could be associated to an increase of attached bacteria [12] in a nutrient deprived medium (like PBS). It is worth mentioning that the DLC coatings have recently been proposed to improve the wear performance of UHMWPE implants [2]. The lower adherence performance of this surface with respect to the uncoated surface represents a preliminary important result.

On the other hand, the incorporation of a small amount of F in the inert DLC structure improves the antiadherent performance of the UHMWPE coated surfaces. Although it is expected that the wear performance of F-DLC coating on UHMWPE be worst, it is also true that only a very thin layer of this compound is necessary to handicap the infection of this surface during surgery.

Superhydrophobic surfaces can be obtained with a combination of high fluorine content C-F polymer, with approximately even composition of all possible C-F bond type functionalities and a surface roughness in the range of tenths of microns. However, this not diminishes bacterial adherence for any of the strains used in this study. Thus, note that not necessarily the most hydrophobic surface presents the lowest adherence performance.

5. Summary and conclusions.

DLC and other fluorinated carbon based functionalities can successfully be incorporated on UHMWPE surfaces at room temperature by RF plasma deposition. Several type and amount of C-F chemical bonds can be added to the UHMWPE surface varying process parameters. Thus, incorporation of increasing amount of Fluorine atoms to an initial DLC structure results in the presence of more oxidized CF_X species. Incorporation of F atoms in the C matrix is correlated to an increase of water contact angle of the corresponding surfaces.

In general, *S. epidermidis* was more adherent than *S. aureus* in all the tested surfaces. *S. aureus* and *S. epidermidis* showed less bacterial adherence on DLC/UHMWPE than the raw UHMWPE, with statistically significant difference. The lowest bacterial adherence was registered for F-DLC/UHMWPE surfaces exposed to *S. aureus*. However, this surface showed similar adherence than the raw UHMWPE material to *S. epidermidis*.

The most hydrophobic surface (CF_X/UHMWPE) showed similar adherence of both bacteria, with the same performance of that obtained for raw UHMWPE (*S. aureus*) and F-DLC/UHMWPE (*S. epidermidis*).

Acknowledgments

We wish to thank the Spanish Ministerio de Educación y Ciencia (project CONSOLIDER FUNCOAT CSD2008-00023) and Junta de Andalucía (project P09-CTS-5189) for their financial support. AT received a research JAE predoc contract from CSIC, and GdP and DM received research contracts from Comunidad de Madrid (CPI/0305/2007) and Fundación Conchita Rábago, respectively.

References.

[1] Katsikogianni M and Missirlis Y F 2004 *European Cells Materials* **8** 37.

[2] Puértolas J A, Martínez-Nogués V, Martínez-Morlanes V, Mariscal M D, Medel F J, López-Santos C and Yubero F 2010 *Wear* **269** 458.

[3] Rico V J, Yubero F, Espinós J P, Cotrino J, González-Elipe A R , Garg D and Henry S, 2007 *Diamond Relat. Mater.* **16** 107.

[4] Valle J, Toledo-Arana A, Berasain C, Ghigo J M, Amorena B, Penadés J R and Lasa I, 2003 *Mol Microbiol* **48** 1075.

[5] Kinnari T J, Soininen A, Esteban J, Zamora N, Alakoski E, Kouri V P, Lappalainen Konttinen Y T, Gómez-Barrena E and TiainenV M 2008 *J Biomed Mat Res Part* A **86** 760.

[6] *ImageJ software* is a freely available (http://rsbweb.nih.gov/ij/index.html) image processing software developed by W. Rasband.

[7] Donlan RM and Costerton JW. 2002 *Clin Microbiol Rev*, **2** 167.

[8] Keogh J R and Easton J W. 1994 *J Lab Clin Med*, **124** 537.

[9] Kinnari T J, Peltonen L I, Kuusela P, Kivilahti J, Könönen M and Jero J 2005 . *Otol Neurotox* **26** 380.

[10] An Y H, Bradley J, Powers D L and Friedman R J. 1997 *J Bone Joint Surg* **5** 816.

[11] An Y H, Blair B K, Martin K L, Friedman R J 2000 Macromolecule surface coating for reventing bacterial adhesion. In *Handbook of Bacterial Adhesion: Principles, Methods, and Applications*. Eds, An YH and Friedman RJ. Totowa, NJ: Humana Press Inc 2000; 609-625.

[12] Donlan R M. 2002 *Emerg Infect Dis*, **8** 881.

Journal of Physics: Conference Series **252** (2010) 012015 doi:10.1088/1742-6596/252/1/012015

Antibacterial properties of biomedical surfaces containing micrometric silver islands

R Pérez-Tanoira[1,*], C Pérez-Jorge[1], J L Endrino[2], E Gómez-Barrena[3], D Horwat[4], J F Pierson[4], J Esteban[1]

[1]Department of Clinical Microbiology, IIS-Foundation Jiménez Díaz, Madrid, Spain
[2] Instituto de Ciencia de Materiales de Madrid, ICMM. Madrid, Spain
[3] Department of Traumatology and Orthopaedic, IIS-Foundation Jiménez Díaz, Madrid, Spain
[4] Institut Jean Lamour, Ecole des Mines de Nancy, Nancy, France

*E-mail: rptanoira@fjd.es

Abstract. A set of Cu-Mn-O and Ag-Cu-Mn-O films were sputter-deposited onto polished Ti-6Al-4V coupons and the microbiological adherence of *Staphylococcus sp.* was studied in these biomedical surfaces modified using advanced ternary and quaternary oxides that incorporated micrometric silver islands. The as-deposited ternary and quaternary compounds were amorphous. Upon air annealing the Ag-Cu-Mn-O films, silver-oxygen bonds in the compound destabilize, resulting in the segregation of metallic silver in the form of micrometric layered silver islands with high specific area dispersed at the surface of the remaining oxide. Silver is well known to have a natural biocidal character and its presence in the surface forming large micrometric escalonated islands is, in principle, predicted to enhance the antimicrobial properties of biomedical surfaces. Microbial adhesion tests were performed in triplicates using collection strains of *Staphylococcus aureus* and *Staphylococcus epidermidis*. Preliminary results indicate that both strains showed decreased adherence to modified materials, *S. epidermidis* showed higher adherence these materials than *S. aureus*, however, there was no statistically significant differences between Cu-Mn-O and Ag-Cu-Mn-O containing silver islands.

1. Introduction

Prosthetic joint implants are used with increased frequency to improve our quality of life. In 2004, 265,441 total hip arthroplasties and 496,018 total knee arthroplasties were performed in U.S. alone. Of these, an estimate of 1.23% total hip arthroplasties and 1.21% total knee arthroplasties were treated for infection. About 8% of hip arthroplasties and 15% of knee arthroplasties revision surgeries were a direct result of infection [1].

The infections due to prosthetic joint implantation have severe consequences not only for patients but also for society because of long hospital stays, long and expensive treatments, and multiple surgeries with severe clinical and economic consequences [2].

Joint prosthetic infection costs about $50,000 U.S. dollars per episode while the associated mortality rate may be as high as 2.5%. In addition, if the infection persists into the deep tissue, amputation may also be required [1].

Although they are less common than infections related to catheters, infections associated with surgical implants are generally more difficult to manage because they require a longer period of antibiotic therapy and repeated surgical procedures [3]. The diagnosis and treatment of prosthetic osteoarticular infections are further complicated by the development of a bacterial biofilm, where the bacteria have changed their phenotypes to an extremely resistant sessile form of life [2,4]. This biofilm is responsible for many chronic infections and prevents proper integration of the implant to the surrounding tissue [1].

Bacterial adhesion to a material surface is the first step in biofilm development. It can be described as a two-phase process including an initial, instantaneous, and reversible physical phase (phase one) and a time-dependent and irreversible molecular and cellular phase (phase two). On the first phase, the planktonic bacteria move to or are moved to a material surface through and by the effects of physical forces, such as Brownian motion, van der Waals attraction forces, gravitational forces, the effect of surface electrostatic charge, and hydrophobic interactions. In the second phase of adhesion, molecular reactions between bacterial surface structures and substrate surfaces become predominant. This implies a firm adhesion of bacteria to a surface by the bridging function of bacterial surface polymeric structures, which include capsules, fimbriae or pili, and slime. In fact, the functional part of these structures should be the adhesins that are parts of these structures, especially when the substrates are host tissues [5]. The adherence process depends largely on the surface and near-surface atomic structure and composition of implanted biomaterials [6]. The factors influencing bacteria adherence to a biomaterial surface include chemical compositions of the material, surface charge, hydrophobicity, and simply surface roughness or physical configuration. Also, their surface energy, empty binding sites, hydrophobic or hydrophilic characteristics can be quickly altered by the adsorption or binding of serum proteins and the formation of biofilms [5]. The number of adherent bacteria that bind to a metal surface is dependent upon the strain of the microorganism and the type of metal [5,6].

The ability to form a biofilm affords at least two properties: the adherence of cells to a surface and multiplication of bacteria to form multilayered cell clusters [7]. Thus, it can be argued that the prevention of bacterial adhesion without drugs may be one of the best ways to reduce orthopedic implant infection [1]. It would be desirable to develop biomedical coatings for implants which are repellent to bacteria to minimize the colonization of the implant surface with circulating planktonic bacteria.

Silver nanoparticles have been shown to possess an unsurpassed antimicrobial spectrum, with efficacy against 150 different pathogens. Silver ions bind strongly to electron donor groups on sulfur-, oxygen- or nitrogen-containing enzymes. These ions displace other cations (e.g., Ca^{2+}) important for enzyme function. In addition, nanocrystalline silver also provides broad-spectrum fungicidal action and only small amounts of silver ion concentrations are required for microbicidal activity [8]. Cu^{2+} and Ag^{+} ions are small enough to disrupt bacterial cell membranes and gain entry in order to disrupt enzyme function. Indirect effects through changes in the surrounding charge environment may also impact on the effectiveness of nanoparticulate metals against microorganisms [9]. These changes can also affect adherence, a process where differences in physical properties could have a high effect in the results.

The reactive sputtering process is well known for its ability to synthesize metastable phases in thin film forms. It is suitable for the deposition of silver-cooper oxides films with various stoichiometries. Since the as-deposited films are amorphous, air annealing has been performed to crystallize the coatings. Annealed films are composed of silver islands exhibited a highly facetted morphology [10]. The annealing of amorphous Ag–Cu–Mn–O films has proven to be an original method to synthesize micrometric silver islands on an oxide surface. The observed high specific area of silver islands seems very promising for applications such as antibacterial coatings. The purpose of this study is to determine whether silver islands can reduce the adhesion of *S. aureus* and *S. epidermis* bacteria.

2. Experimental

2.1 Materials and methods

The ternary (and quaternary) films were deposited onto Ti-6Al-4V substrate by co-sputtering of manganese and copper (and $Ag_{60}Cu_{40}$) targets in a reactive Ar–O_2 reactive mixture. The currents applied to the Mn and the Ag/Cu targets were adjusted to deposit films with an approximate atomic ratio (Ag+Cu)/Mn of 2.9. The as-deposited films were X-ray amorphous and air annealing in the 350–550 ºC range of the Ag-Cu-Mn-O films did not allow the crystallization of the delafossite (Ag_2CuMnO_4) phase. On the other hand, such treatments induced the formation of visible silver islands located on the surface of the film, as can be observed in the photograph of Fig. 2. The morphological properties of the silver containing films were characterized using a perfilometer instrument (Veeco Dektak 150) and Vision 3-D analysis software. This technique allowed obtaining a three-dimensional representative image (0.5 mm x 0.5 mm) of the surface of the prepared silver islands.

2.2 Staphylococcal adhesion experiments

Ti-6Al-4V coupons coated with Ag-Cu-Mn-O films containing islands of silver located on the surface of the film (Fig. 1c and Fig. 2) were compared against controls of uncoated Ti-6Al-4V coupons (Fig. 1a) and Ti-6Al-4V coupons coated with an amorphous Cu-Mn-O film (Fig. 1b).

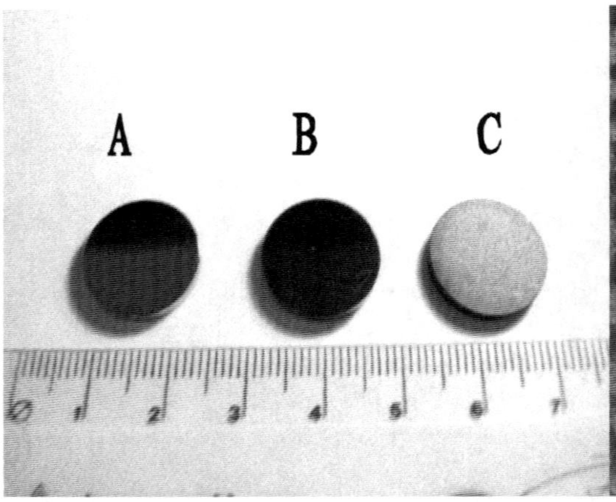

Figure 1. Photography of the coupons. a) Ti-6Al-4V coupons, b) Ti-6Al-4V coupons coated with Cu-Mn-O and c) Ti-6Al-4V coupons coated with annealed Ag-Cu-Mn-O

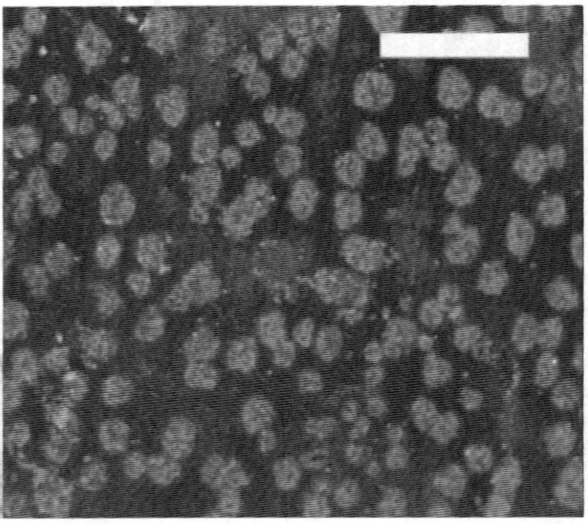

Figure 2. Photography of the surface of the Ti-6Al-4V coupons coated with annealed Ag-Cu-Mn-O. The white bar represents 500 µm length.

Staphylococci (*Staphylococcus aureus* and coagulase-negative *Staphylococcus* species) are the most frequent infective agents, followed by streptococci, enterococci, gram-negative bacilli, and *Propionibacterium acnes* [4]. *S. aureus* and *S. epidermidis* were selected for testing of bacterial adhesion. This reflects the potential of the soluble planktonic bacteria to stick to the implant surface and to colonize it. In this study the adhesion is tested using collection strains, which are reproducible and available for all investigators.

Staphylococcal adhesion experiments were performed as described by Kinnari et al. Briefly the biofilm-forming collection strains *S. aureus* 15981 [11] and *S. epidermidis* ATCC 35984 were cultured overnight in tryptic soy broth (bioMérieux, Marcy l'Etoile, France) at +37ºC. After culture, bacteria were harvested by 10-min centrifugation at 3500 x g at room temperature. Supernatant was discarded and the pellet was washed three times with sterile phosphate buffered saline (PBS). Bacteria were then suspended and diluted in PBS to 10^8 colony-forming units (CFU)/mL. The biomaterial discs were placed into the bacterial suspension and incubated during 90 min at +37ºC. Afterwards, the biomaterial plates were rinsed three times with sterile PBS to remove any nonadherent bacteria [12].

Dried plates were stained for 15 min with Backlight© Live/Dead stain. On each plate, 8 fields were viewed and photographed with Nikon Coolpix 8400 (Nikon, Melville, NY) under a fluorescence microscope at 40x magnification (Figure 5 and figure 6). All experiments were performed in triplicates.The number of microphotographs studied was 24 per each material and bacterium. The surface area covered with adhered bacteria was calculated using the ImageJ software (National Institue of Health, Bethesda, MD)

3. Results and discussion

3.1 Thin film morphology

X-ray diffraction (XRD) patterns (not shown) revealed that the as-deposited coatings were amorphous and there were only peaks for crystalline fcc-Ag peaks in the case of the annealed Ag-Cu-Mn-O film due to the formation of the silver islands. 3D perfilometer plot of the annealed Ag-Cu-Mn-O coating showed that about one fourth of the surface is covered by the Ag islands (Fig. 3). Each individual silver island appears to be composed of about hundred silver rounded agglomerates with a thickness of over 1μm, which is larger than the as-deposited film thickness (which was about 0.75 μm). Silver agglomerates around the perimeter of the islands had higher thickness than agglomerate grains within the island. The overall increase in surface determined by the surface index, calculated using Vision program (Veeco), was 1.066. This means that the overall increase in the lateral surface area was only about 0.66%. Observation of one individual silver island at higher magnification is shown in Fig. 4, together with the surface area and surface index calculations for this region. The slight increase in surface index close to 1.01 indicates that the surface of the silver islands is very flat and they contained no faceted morphology.

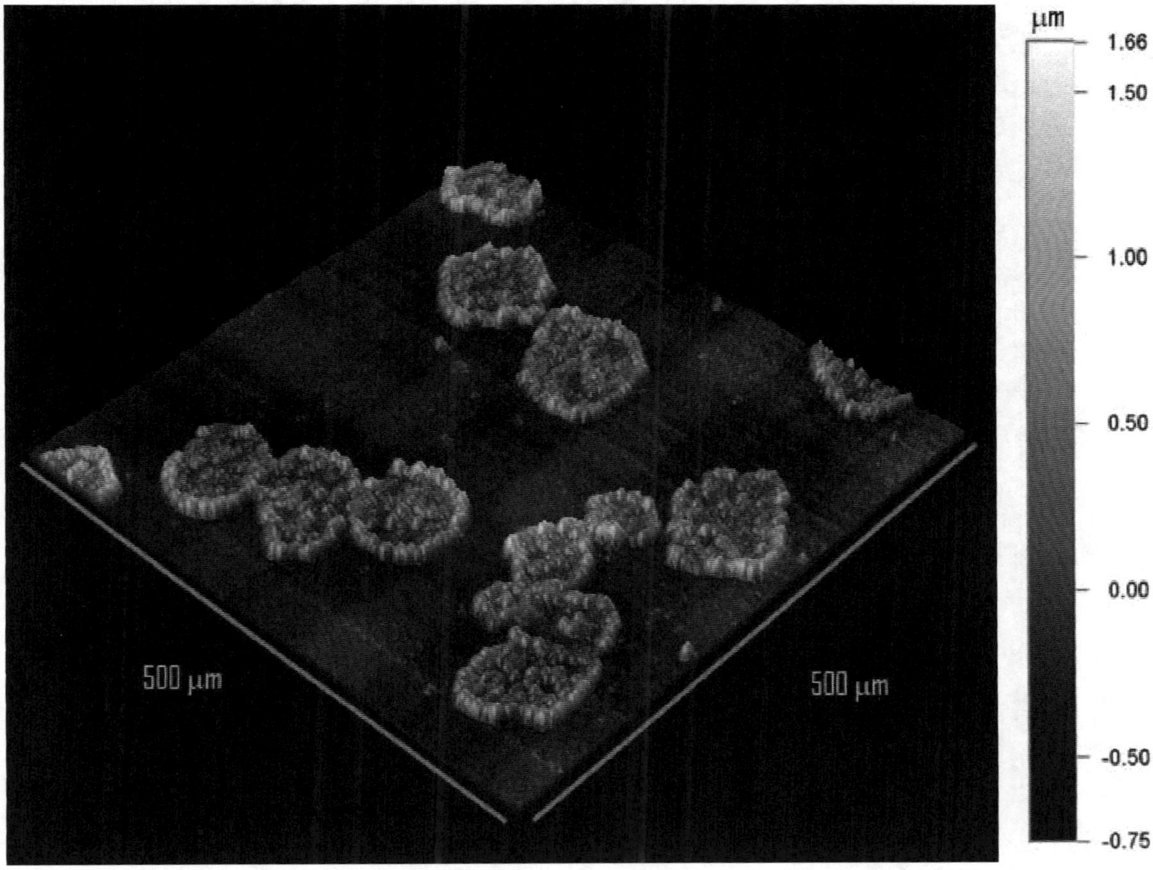

Figure 3. 3-D plot of a perfilometer surface scan (500μmx500μm) of annealed Ag-Cu-Mn-O film deposited onto a Ti6Al4V. Silver islands on the surface are formed due to annealing at 550°C for 2 hours

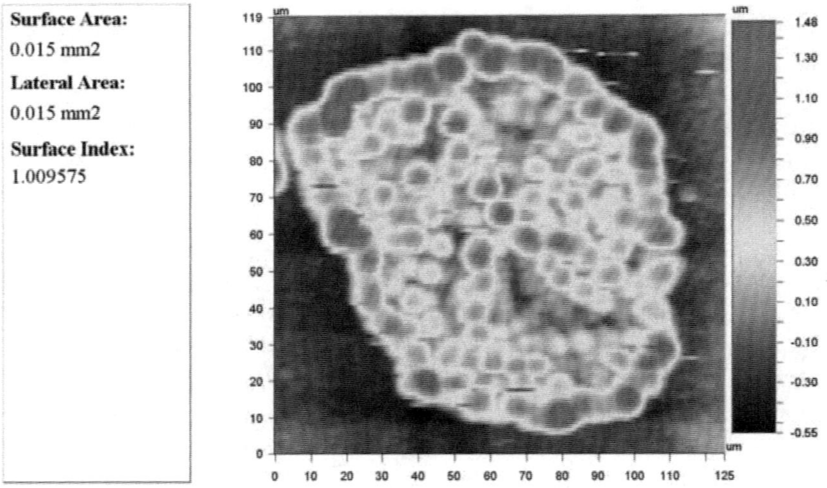

Figure 4. Coloured contour plot and lateral area analysis of perfilometer surface scan of one silver island.

3.2 Bacterial adhesion tests

The adherence results, represented in the table 1 and in the figure 5, indicate that both strains showed lowered adherence to modified surface materials.

Table 1: Percentage of Ti-6Al-4V surface, Ti-6Al-4V with Ag-Cu-Mn-O surface and Ti-6Al-4V with Cu-Mn-O surface covered by *S. aureus* or *S. epidermidis.*

Bacteria	Material	Percentage of biomaterial surface covered (Mean ± Std Dev)
S. aureus	Ti-6Al-4V	3,1167 (±0,6651)
	Ag-Cu-Mn-O	1,6750 (±0,3698)
	Cu-Mn-O	1,7583 (±0,3374)
S. epidermidis	Ti-6Al-4V	5,0542 (±1,2779)
	Ag-Cu-Mn-O	3,2833 (±1,6056)
	Cu-Mn-O	2,9208 (±1,4788)

Figure 5. Mean percentage of biomaterial surface covered with each bacteria and for each material**.**

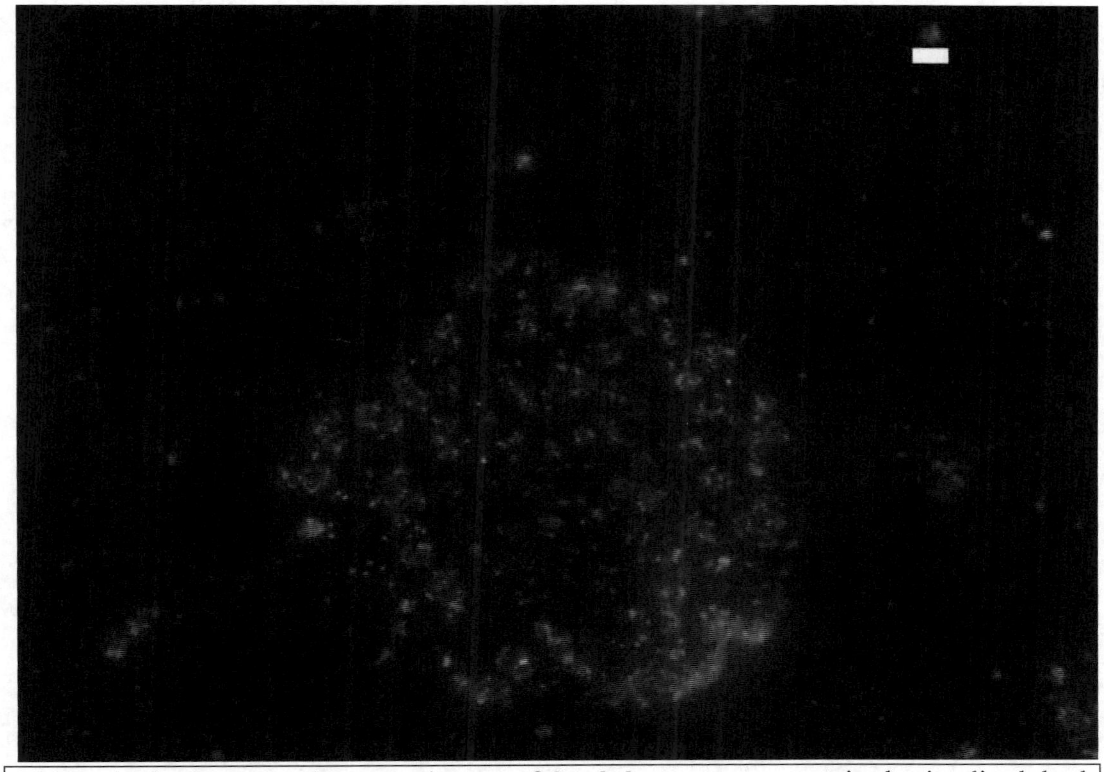

Figure 6. Fluorescence microscope images of *Staphylococcus aureus* stained using lived-dead strain attached to the Ti-6Al-4V coated with annealed Ag-Cu-Mn-O. The white bar represents 10 μm length.

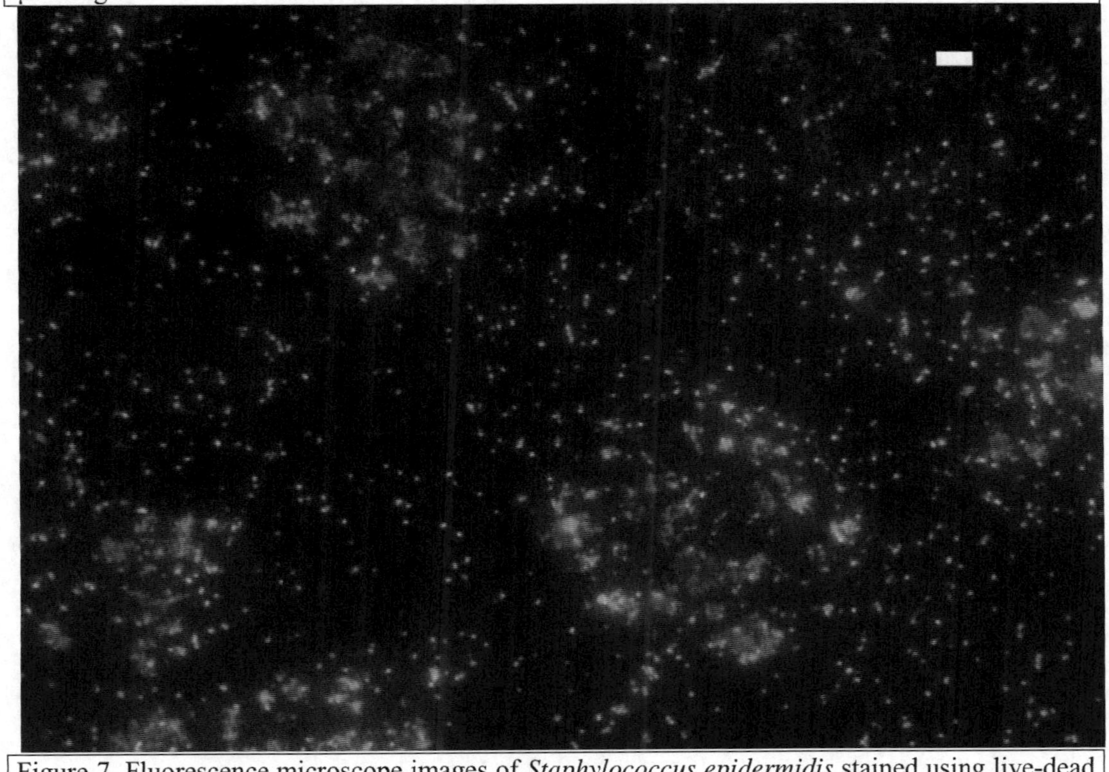

Figure 7. Fluorescence microscope images of *Staphylococcus epidermidis* stained using live-dead strain attached to the Ti-6Al-4V coated with annealed Ag-Cu-Mn-O. The white bar represents 10 μm length.

The surface of area covered by adhered bacteria was adjusted to the surface available for bacterial cells in case of the discs with silver islands. These islands represent an increase of 0.1 % in this surface compared with other materials due to the morphology of the silver islands. We have also detected some fluorescence of the silver islands. All calculations have been performed take into account this fact. Figures 6 and 7 show the fluorescence microscope images corresponding to the sample containing silver islands for *S.aureus* and *S.epidermidis,* respectively. These images of the adhered bacteria show different taint when they are on the top of islands of silver. Moreover, the dead bacteria which are tainted in orange-red appear to describe the morphology of smaller silver grains within the islands, as described in Fig. 3 and 4.

3.3 Statistical analysis

For the statistical study was using not parametric test. Mann-Whitney/Wilconxon test was using for two samples and Kruskal-Wallis test was using for more than two samples.

The differences between Ti-6Al-4V surface and Ti-6Al-4V with Ag-Cu-Mn-O were statiscally significant for *S. aureus* ($p<0.0001$, test of Kruskal-Wallis) and *S. epidermidis* ($p= 0.0005$, test of Kruskal-Wallis).

The differences between Ti-6Al-4V surface and Ti-6Al-4V with Cu-Mn-O were statiscally significant for *S. aureus* ($p<0.0001$, Test of Kruskal-Wallis) and *S. epidermidis* ($p<0.0001$, test of Kruskal-Wallis).

There were no statistical differences between Ti-6Al-4V with Cu-Mn-O and Ti-6Al-4V containing silver islands for *S. aureus* ($p=0.3188$, Test of Kruskal-Wallis) and *S. epidermidis* ($p=0.5493$, test of Kruskal-Wallis), even when surface was adjusted to the higher exposed surface area of the silver-containing compounds. Such increase of surface was calculated to be 0.1% higher than other materials.

When we compared the adhesion of both strains, the difference between the adhesion of *S. aureus* and *S. epidermidis* was statistically significant for Ti-6Al-4V ($p< 0.0001$, test of Kruskal-Wallis), for Ti-6Al-4V with Ag-Cu-Mn-O ($p=0.0014$, test of Kruskal-Wallis) and for Ti-6Al-4V with Cu-Mn-O ($p=0.0109$, test of Kruskal-Wallis)

2. Conclusions and future work

Biomedical Ti-6Al-4V plates were successfully coated using amorphous Ag-Cu-Mn-O or Cu-Mn-O films deposited using the reactive magnetron sputtering technique. Films composed of Ag-Cu-Mn-O were thermally treated and upon annealing silver islands that contained smaller silver agglomerates were formed at the film surface. Coated plated had a decreased bacterial adherence of collection strains of *S. aureus and S. epidermidis* in comparison to uncoated plates. However, differences were not detected between both surface coatings with these strains. *S. epidermidis* showed higher adherence for all materials than *S. aureus*.

This study evaluated the rough adherence of collection strains. These strains have lower genetic load than clinical strains isolated of patients because they are laboratory-adapted strains which loss genes due to several passages on culture medium. Because this fact, it must be interesting to prove these materials with strains isolated of patients, which can behave in a different way than collection strains due to their different pathogenic properties. Further experiments are currently being performed to evaluate such adherence using clinical strains isolated from patients with prosthetic joint infections (which can show different behavior than laboratory-adapted strains). These experiments also study the antibacterial effect of silver through evaluation of percentages of live and dead bacteria present in the surfaces using the live/dead strain.

Future studies must analyze the effect of silver islands on biofilm development, which needs longer incubation times and where silver elution can have a prolonged effect on sessile cells, and effect that cannot be detected in adhesion experiments, where only immediate interactions between bacteria and the surface can be studied.

Acknowledgements
This study was realized thanks to a help of the program CONSOLIDER-INGENIO 2010 FUNCOAT-CSD2008-00023. JLE also acknowledges Spanish MICINN Project FIS2009-12964-C05-04

References

[1] Puckett S, Taylor E, Raimondo T, Webster T. 2010 The relationship between the nanostrucutre of titanium surfaces and bacterial attachment. *Elsevier* Biomaterials 31 706-713

[2] Esteban J, Gomez-Barrena E, Cordero J, Zamora N, Kinnari T and Fernandez R. 2008 Evaluation of Quantitative Analysis of Cultures from Sonicated Retrieved Orthopedic Implants in Diagnosis of Orthopedic Infection Journal of Clinical Microbiology Vol.46 No. 2 p 488-492

[3] Rabih O, Darouiche. 2004 Treatment of infections associated with surgical implants N ENGL J MED 350:1422-9

[4] Trampuz A, Zimmerli W Diagnosis and Treatment of Implant-Associated Septic Arthritis and Osteomyelitis. 2008 *Current Infectious Disease Reports* 10: 394-403

[5] An YH, Friedman RJ. 1998 Concise review of mechanisms of bacterial adhesion to biomaterial surfaces Journal of Biomedical Materials Research 43:338-348

[6] Gristina A. Biomaterial-centered infection:Microbial adhesion versus tissue integration. Science 1987;237:1588-1595

[7] Götz F. 2002 *Staphylococcus* and biofilms Molecular Microbiology 43(6), 1367-1378

[8] Narayan RJ, Abernathy H, Riester L, Berry CJ and Brigmon R. 2005 Antimicrobial Properties of Diamond-like Carbon-Silver-Platinum Nanocomposite Thin Films Journal of Materials Engineering and Performance Volume 14(4)-435

[9] Ren G, Hu D, Cheng E, Vargas-Reus M, Reip P, Allaker Robert. 2009 Characterisation of copper oxide nanoparticles for antimicrobial applications. International Journal of Antimicrobial Agents 33(2009)587-590

[10] Pierson JF, Baija M, Horwat D. 2009 Silver islands formed after air annealing of amorphous Ag-Cu-Mn-O sputtered films. Journal of Crystal Growth 311(2009) 349-354

[11] Valle J, Toledo-Arana A, Berasain C, Ghigo JM, Amorena B, Penades JR, Lasa I. SarA and not SigB is essential for biofilm development by *Staphylococcus aureus*. Molecular Microbiology 2003; 48:1075–1087

[12] Kinnari T, Soininen A, Esteban J, Zamora N, Alakoski E. 2007 Adhesion of staphylococcal and Caco-2 cells on diamond-like carbon polymer hybrid coating Journal of Biomedical Materials Research 86A:760-768

Evaluation of the effect of vitamin E doped UHMWPE on biofilm development and infection using an *in vivo* experimental model

Molina-Manso D[1], Gómez-Barrena E[2], Alonso-Rodríguez N[1], Sandoval E[2], Cordero J[4], Fernández-Roblas R[1], Puértolas JA[3], Esteban J[1].

[1]Dept. of Clinical Microbiology, IIS Fundación Jiménez Díaz

[2]Dept. of Orthopaedics, Hospital Fundación Jiménez Díaz, Autónoma University, Madrid, Spain

[3]Dept. of Material Science, Zaragoza University, Spain

[4]Dept. of Orthopaedics, Hospital La Princesa, Autónoma University, Madrid, Spain

E-mail: dmolina@fjd.es

Abstract. The aim of our study was the evaluation of an in vivo experimental model of implant-related septic arthritis. 5 x 10 mm strips out of a polyethylene sheet 500 microns thick were incubated with a 0.5 McFarland bacterial suspension of collection strains *S. aureus* and *S. epidermidis*. A surgical experimental model was prepared, implanting the samples in the subquadricipital articular space of the rabbit's knee. After 7 days of survival, the knee joint was opened wide through the previous surgical approach and the polyethylene sample was retrieved. The UHMWPE samples were processed following a sonication and quantification protocol. Ten rabbits for each species were studied, five with each material. Negative controls (UHMWPE strips without attached bacteria) were also implanted in contralateral knees. *S. aureus* colonies were detected only in three rabbits with non-treated UHMWPE and in two with vitamin E-doped UHMWPE. No differences in colony counts were observed for *S. aureus*. No growth was detected for *S. epidermidis*, although clinical signs of infection were detected in all animals with inoculated samples. The model was useful to evaluate the effect of modifications in biomaterials, although highly pathogenic bacteria are needed to obtain quantifiable data.

1. Introduction

Bacterial adherence to biomaterials is the origin of implant related infection, but limited information is available about this type of infections in experimental models.

Ultra-high molecular weight polyethylene (UHMWPE) is widely used in osteoarticular prostheses, particularly in knee and hip prostheses. Some studies demonstrate that UHMWPE is altered by the addition of Vitamin E (VE) to decrease material oxidation [1-5]. Potential effects of VE on bacterial adherence are currently being investigated, and the aim of this study was to evaluate the effect of UHMWPE-VE on biofilm development and infection in an *in vivo* experimental model with rabbits.

2. Materials and methods

2.1. Collection strains and UHMWPE
S. aureus 15981 provided by Dr. Lasa [6] and *S. epidermidis* ATCC 35984 were used. The investigated material samples were 5 x 10 mm strips obtained from a UHMWPE virgin and commercial VE doped sheet 500 microns thick.

2.2. UHMWPE samples preparation
After overnight culture in Tryptic-soy broth, bacteria were harvested after 20 minute centrifugation at 3500g, and washed twice with sterile Phosphate Buffered Saline (PBS). The bacteria were then suspended and diluted in PBS to a concentration of 10^8 colony-forming units (CFU)/ml. UHMWPE samples were placed in this bacterial suspension and incubated for 90 minutes at +37°C. After the incubation, specimens were rinsed twice with PBS proceeding with implantation.

2.3. Experimental model
A surgical experimental model was prepared, implanting the samples in the subquadricipital articular space of the rabbit's knee as follows.

Under general anesthesia with IM Ketolar, a medial proximal parapatelar approach was performed, incising up to the quadricipital tendon. A blunt instrument was introduced from distal to proximal and, while maintaining the instrument as a guide, the sample was introduced and the joint was closed.

After 7 days the rabbits were slaughtered by penthobarbital overdose. Following the previous approach, the knee joint was opened wide and the polyethylene sample was retrieved with sterile forceps, and then introduced into a tube with a known amount of sterile PBS.

Samples were sonicated during 5 minutes and the number of bacteria was quantified by 1: 10 serial plate counts

Ten rabbits for each specie were studied, five with each material. Negative controls (UHMWPE strips without attached bacteria) were also used. Figure 1 shows the implantation process.

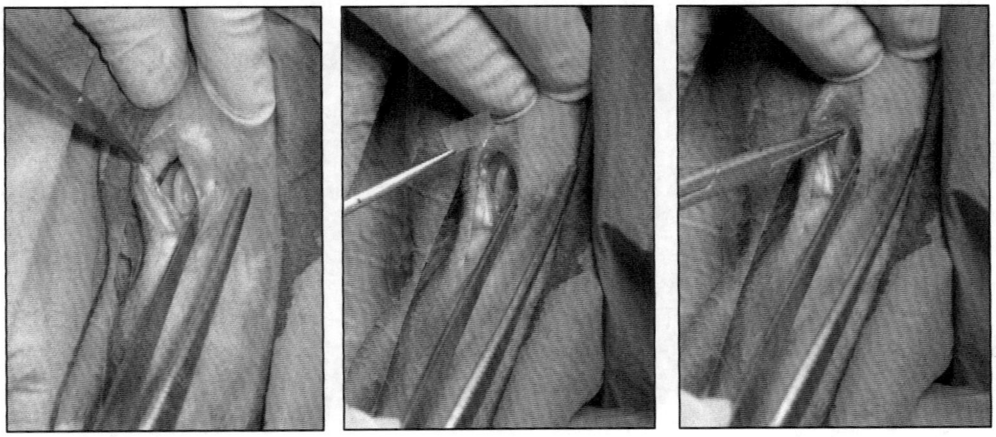

a) b) c)

Figure 1: Images of the arthrotomy. a) Opening of the articular space. b) UHMWPE sample.

c) Introduction of the sample in the knee

This protocol was repeated for *S. epidermidis* strain with the following modification: rabbits were slaughtered after 3, 7 and 14 days. After these periods, UHMWPE samples were processed as described above.

3. Results

CFU´s were only found in 5 out of 20 rabbits, all of them from the specie *S. aureus*, 3 in rabbits with non-treated UHMWPE and 2 with vitamin E doped UHMWPE. No differences in colony counts were observed for *S. aureus*.

No growth was detected for *S. epidermidis* in any experiment, although clinical signs of infection were detected in all animals with inoculated samples. Table 1 shows the colony counts results.

Table 1. Colony counts of UHMWPE with *S. aureus* after extraction from the rabbit's knee

NUMBER UHMWPE	SONICATE DILUTIONS 10^1	10^2	10^3	CFU/ mL	TOTAL CFU
1 V	10	1	X	1×10^2	3×10^2
2 V	58	4	3	$4{,}9 \times 10^3$	$1{,}5 \times 10^4$
3 V	10	1	X	1×10^3	3×10^3
6 VE	77	8	2	$7{,}8 \times 10^3$	$2{,}3 \times 10^4$
10 VE	202	36	6	$2{,}8 \times 10^4$	$8{,}4 \times 10^4$

4. Discussion and Conclusions

Investigation of bacterial adherence related to implanted materials is a matter of increasing frequency and interest. *S. epidermidis* and *S. aureus* are the most frequent microorganisms involved in biomaterial-related infection, that always occur in relation to biofilm formation. This structure constitutes a protected mode for the bacteria to survive in a hostile environment.

Some factors, such as the surface of the implant, the local damage of the tissues when the devices are implanted or the systemic general condition of the individual may affect to the bacterial adhesion and thus, to the infection development [7]. For this reason it is important to realize more studies testing different kind of biomaterials.

A new experimental model was developed and evaluated that enables in-vivo testing of bacterial adherence to prosthetic materials [8]. *In vivo* studies are reported to simulate biomaterial infection. Even if the need to simulate human clinical conditions suggests using large animals, rabbit models are those most frequently found in the literature [7], because of their adequate size and fast responses.

Many strategies have been developed to prevent or reduce bacterial contamination in biomaterials. These include morphological or physical modifications. In this study, we use an *in vivo* model to test the effect of VE in UHMWPE. The model was useful to evaluate the effect of modifications in biomaterials, although highly pathogenic bacteria are needed to obtain quantifiable data. The absence of quantificable results with *S. epidermidis* could be due to the differences between immune systems of humans and rabbits, even when a biofilm-producing strain was used and despite the fact of presence of clinical infection in the animals.

S. aureus consistently produced quantifiable joint infection when seeded in UHMWPE, but *S. epidermidis* produced clinical infection signs without objective, quantitative culture despite high inoculii and clinically aggressive strains.

Acknowledgements

The work is funded by a grant from MEC (Spain), MAT2006-12603-CO2-O2. and by MCYT (CONSOLIDER-INGENIO 2010 CSD 2008-0023 FUNCOAT). DMM was granted by Fundación Conchita Rabago de Jimenez Diaz.

References

[1] Parth M, Aust N, Lederer K. 2002. Studies on the effect of electron beam radiation on the molecular structure of ultra-high molecular weight polyethylene under the influence of alpha-tocopherol with respect to its application in medical implants. *J Mater Sci Mater Med.* **13**(10):917-21.

[2] Wolf C, Krivec T, Blassnig J, Lederer K, Schneider W. 2002. Examination of the suitability of alpha-tocopherol as a stabilizer for ultra-high molecular weight polyethylene used for articulating surfaces in joint endoprostheses. *J Mater Sci Mater Med.* **13**(2):185-9.

[3] Oral E, Greenbaum ES, Malhi AS, Harris WH, Muratoglu OK. 2005. Characterization of irradiated blends of alpha-tocopherol and UHMWPE. *Biomaterials.* **26**(33):6657-63.

[4] Shibata N, Tomita N, Onmori N, Kato K, Ikeuchi K. 2003. Defect initiation at subsurface grain boundary as a precursor of delamination in ultrahigh molecular weight polyethylene. *J Biomed Mater Res A.* **67**(1):276-84.

[5] Oral E, Wannomae KK, Hawkins N, Harris WH, Muratoglu OK. 2004. Alpha-tocopherol-doped irradiated UHMWPE for high fatigue resistance and low wear. *Biomaterials.* **24**:5515-22.

[6] Valle J, Toledo-Arana A, Berasain C, Ghigo JM, Amorena B, Penades JR, et al. 2003. SarA and not sigmaB is essential for biofilm development by *Staphylococcus aureus. Mol Microbiol.* **48**(4):1075-87

[7] Rimondini L, Fini M, Giardino R. 2005. The microbial infection of biomaterials: a challenge for clinicians and researchers. A short review. *J Appl Biomater Biomech.* **3**(1): 1-10

[8] Eichinger WB, Goppel G, Mendler N, Mattes J, Lankes E, Botzenhardt F, Bauernschmitt R, Lange R. 2002. In-vivo bacterial adherence to intracardiac prosthetic materials: a new experimental model. *J Heart Valve Dis.* **11**(3):438-46.

Si+ ion implantation reduces the bacterial accumulation on the Ti6Al4V surface

A.M. Gallardo-Moreno[1,3], M.A. Pacha-Olivenza[3], J. Perera-Núñez[1,3], J.L. González-Carrasco[2,4], M.L. González-Martín[1,3] *

[1]Departamento de Física Aplicada, Facultad de Ciencias, Universidad de Extremadura; [2]Departamento de Metalurgia Física, CENIM (CSIC), Madrid, Spain; CIBER-BBN, Centro de Investigación Biomédica en Red en Bioingeniería, Biomateriales y Nanomedicina, [3]Badajoz, [4]Madrid, Spain

* mlglez@unex.es

Abstract. Ti6Al4V is one of the most commonly used biomaterials in orthopedic applications due to its interesting mechanical properties and reasonable biocompatibility. Nevertheless, after the implantation, microbial adhesion to its surface can provoke severe health problems associated to the development of biofilms and subsequent infectious processes. This work shows a modification of the Ti6Al4V surface by Si+ ion implantation which reduces the bacterial accumulation under shear forces. Results have shown that the number of bacteria remaining on the surface at the end of the adhesion experiments decreased for silicon-treated surface. In general, the new surface also behaved as less adhesive under in vitro flow conditions. Since no changes are observed in the electrical characteristics between the control and implanted samples, differences are likely related to small changes observed in hydrophobicity.

1. Introduction

Bacteria prefer a community-based surface-bound, sedentary lifestyle to a nomadic existence. There may be an obvious explanation for bacterial adhesion, because nutrients in an aqueous environment tend to concentrate near a solid surface [1].

As an oversimplified rule of thumb, primary adhesion between bacteria and abiotic surfaces is generally mediated by nonspecific interactions, whereas adhesion to living or devitalized tissue is accomplished through specific molecular (lectin, ligand or adhesion) docking mechanisms. Primary adhesion is reversible and defines the further adhesion between the bacterial cell surface and the conditioned surface of interest. Once the microorganisms reach critical proximity to the surface, the final determination of adhesion depends on the net sum of attractive or repulsive forces generated between both surfaces, including electrostatic and hydrophobic interactions [2].

The ability to control microbial adhesion is of enormous importance in healthcare, particularly in modern surgery where postoperative implant-associated infections are still an unresolved and serious complication. This challenge is compounded by the ever-increasing problem of antibiotic resistant hospital and community acquired infections [3, 4]. For this reason new strategies are focused on minimizing microbial surface colonization by modifying the biomaterial surface and antibacterial coatings of Ti6Al4V have been recently proposed by different research groups [5-7].

In this line, this work shows a new modification of the Ti6Al4V surface by Si+ ion implantation which reduces the bacterial accumulation under in vitro flow conditions. Surface characterization is carried out by means of hydrophobicity and isoelectric point.

2. Materials and methods

2.1. Ti6Al4V surface modification.
Ion implantation was carried out by using a F_4Si precursor on the surface of 20 mm diameter disks of Ti6Al4V kindly supplied by Surgival SA, Spain. Ti6Al4V without implantation will be denoted as "control" and Ti6Al4V implanted will be denoted as "implanted".

2.2. Surface characterization.
2.2.1. *Hydrophobicity*: It was quantified through the water contact angle (θ_W) on the sample surface.
2.2.2. *Surface Gibbs Energy*: It was evaluated from the contact angles of water, formamide (θ_F) and diiodomethane (θ_D) by applying the Young Equation [8] and Van Oss approach [9]. The surface Gibbs energy (γ^{Total}) was the sum of the Lifshitz-van der Waals component (γ^{LW}) and acid-base component (γ^{AB}) which, in turn, is the geometric mean of the electron-donor (γ^-) and electron-acceptor (γ^+) parameters.
2.2.3. *Isoelectric point*: zeta potential (ζ) was measured as pH function and isoelectric point (IEP) was obtained as the pH at which ζ was zero [10].

2.3. Bacterial strains.
Three gram-positive strains with different EPS-production were used: *Staphylococcus aureus* ATCC29213 (*S. aureus*), *Staphylococcus epidermidis* ATCC35984 (*S. epidermidis*4), *Staphylococcus epidermidis* HAM892 (*S. epidermidis*2).

2.4. Adhesion experiments.
Bacterial adhesion experiments were carried out at 37 °C in a parallel plate flow chamber and the results were analysed in terms of flow or static conditions by selecting a constant flow rate or stopping the bacterial flow. Different protocols were followed:
2.4.1. *Dynamic adhesion*: the flow chamber was placed so that the alloy surface was on the top side of the flow channel and the bacteria were allowed to attach while flowing at a constant rate for 20 min. The number of bacteria at the end the experiments dynamic was denoted by $N_{D\text{-}20min}$. It was also analysed the initial adhesion rate to the surface (j_D).
2.4.2. *Static adhesion*: the sample was placed on the bottom side of the flow channel, the flow was stopped and the bacteria were left to deposit on the surface for different time interval periods. Static adhesion rates (j_S) were obtained and compared to that of dynamic.
2.4.3. *Shear forces*: At the end of the experiments two consecutive air-liquid interfaces were passed through the flow channel and the number of bacteria remaining on the surface were analysed in order to check the strength of the bacterial retention.

3. Results and discussion

Table 1 shows the surface characterization of both control and implanted samples. There are small changes in contact angles but the implanted surface is slightly more hydrophilic than the control. This also implies a higher polarity of the treated surface since the polar liquids (W, F) present lower contact angles and γ^{AB} is higher. There is no difference between samples in the IEP within experimental error.

Figure 1 presents interesting information from the bacterial adhesion tests. It is observed that initial adhesion rates to control and implanted are not statistically different (Fig. 1a). Only in the case of *S. epidermidis2* a decrease in j_D is observed in the implanted sample respect to control. This strain also presents the highest decrease in the final number of adhered bacteria after dynamic experiments (Fig. 1b). $N_{D\text{-}20min}$ also diminishes for *S. aureus* in the implanted. Information provided by static experiments are different to that of dynamic, as showed when comparing Figs 1a and 1c. j_S is always higher than j_D, indicating that static adhesion is highly influenced by sedimentation processes. Relationships between control and implanted in both experiments are also different. In the case of static adhesion Si+ ion implantation never reduces adhesion (Fig. 1c) on the contrary; it is increased for both strains of *S. epidermidis*. An interesting observation is that after the passage of two air-liquid interfaces the bacterial detachment is more effective in silicon-treated surface than in control (Fig. 1d), which could enhance the applicability of this technique in those orthopaedic applications in which shear forces are present.

Table 1. Contact angles of water (θ_W), formamide (θ_F) and diiodomethane (θ_D), Lifshitz-Van der Waals (γ^{LW}) and acid-base (γ^{AB}) components as well as electron-acceptor (γ^+) and electron-donor (γ^-) parameters and the total surface Gibbs energy (γ^{Total}) for control and implanted samples. Isoelectric points (IEP) are also represented.

	Contact Angle			**Surface Gibbs Energy**					**IEP**
	θ_W	θ_F	θ_D	γ^{LW}	γ^{AB}	γ^+	γ^-	γ^{Total}	
Control	58.6 ±1.8	48.7 ±1.6	40.6 ±1.4	32.4 ±0.7	8.1 ±1.7	0.7 ±0.2	24 ±3	40.5 ±2.4	5.5 ±0.9
Implanted	53.5 ±0.7	42.5 ±0.7	38.3 ±0.7	33.1 ±0.6	10.9 ±0.8	1.14 ±0.11	26.1 ±1.1	44.0 ±1.4	4.8 ±0.8

Figure 1. Bacterial adhesion experiments. j_D: initial dynamic adhesion rate locating the sample at the top of the flow channel (a). $N_{D\text{-}20min}$: number of adhered bacteria at the end of the dynamic adhesion experiment (b). j_S: static adhesion rate locating the sample at the bottom of the flow channel (c). Percentage of remaining bacteria after passing two air-liquid interfaces (d).

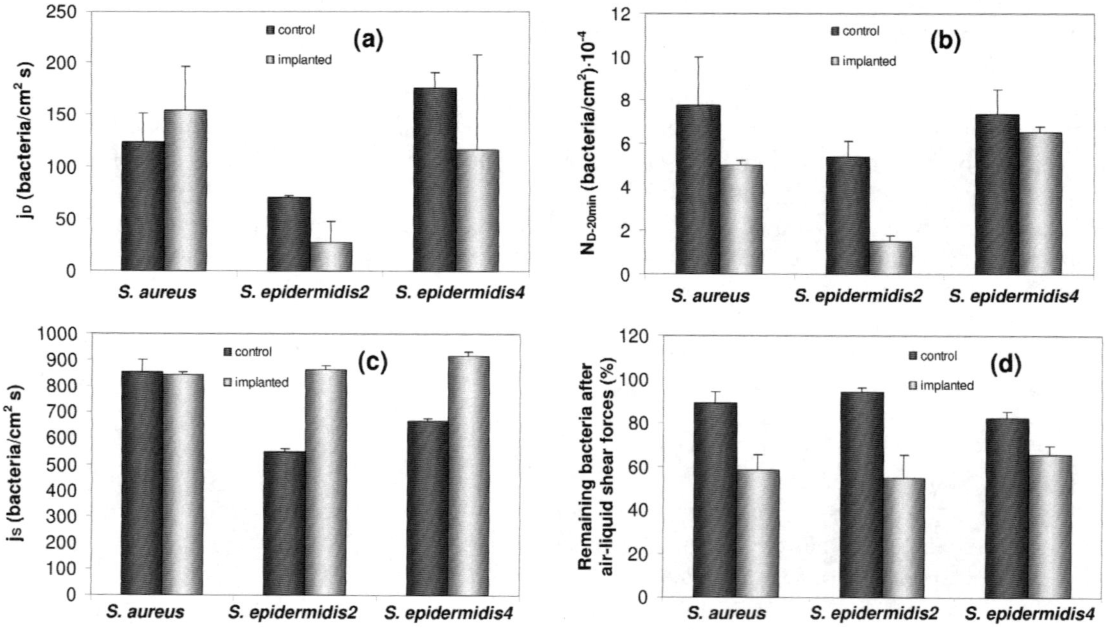

Relationships between physico-chemical surface parameters and bacterial adhesion have been extensively studied [2, 7, 11]. However it is difficult to establish the exact contribution of each magnitude to any particular adhesion process involving microorganisms. Electrostatic interactions tend to favor repulsion, because most bacteria and inert surfaces are negatively charged. *Stenotrophomonas maltophilia* is one exception to this rule, and the overall positive surface charge of this organism, at physiological pH can promote primary adhesion to negatively charged materials such as Teflon [12]. In our case, the IEP obtained for both surfaces indicates that they are negatively charged at physiological pH (about pH=7). As indicated by different research groups, hydrophobic-hydrophilic interactions probably have greater influence on the outcome of primary adhesion [7, 13, 14]. In the present research the small changes in the hydrophobicity of the surface after implantation seems to be related to the lower retention of the three strains (Fig. 1d). Silicon-treated surface become more hydrophilic and this means a higher affinity for water under shear conditions, reducing the final bacterial accumulation on the surface.

Acknoledgments

This work was supported by Grant MAT2009-14695-C04-(01-04) from the "Ministerio de Ciencia e Innovación" and Grant PRI08A124 from the "Consejería de Economía, Comercio e Innovación" of the "Junta de Extremadura".

References

[1] Dunne W M Jr 2002 Bacterial adhesion: seen any good biofilms lately? *Clin. Microbiol. Rev.* **15** 155-166.

[2] An Y H and Friedman R J (ed) 2000 Handbook of bacterial adhesion: principles, methods and applications. Humana Press, Totowa, NJ.

[3] Gould I M 2006 Costs of hospital-acquired methicillin-resistant *Staphylococcus aureus* (MRSA) and its control. *Int. J. Antimicrob. Ag.* **28** 379-384.

[4] Moellering R C Jr 2006 The growing menace of community-acquired methicillin-resistant *Staphylococcus aureus*. *Ann. Intern. Med.* **144** 368-370.

[5] Heidenau F, Mittelmeier W, Detsch R, Haenle M, Stenzel F, Ziegler G and Gollwitzer H 2005 A novel antibacterial titania coating: metal ion toxicity and in vitro surface colonization. *J. Mater. Sci.* **16** 883-888.

[6] Sarró M I, Moreno D A, Ranninger C, King E and Ruiz J 2006 Influence of gas nitriding of Ti6Al4V alloy at high temperature on the adhesion of *Staphylococcus aureus*. *Surf. Coat. Technol.* **201** 2807-2812.

[7] Gallardo-Moreno A M, Pacha-Olivenza M A, Saldaña L, Pérez-Giraldo C, Bruque J M, Vilaboa N and González-Martín M L 2009 *In vitro* biocompatibility and bacterial adhesion of physico-chemically modified Ti6Al4V surface by means of UV irradiation. *Acta Biomater.* **5** 181-192.

[8] Young T 1805 An essay on the cohesion of fluids. Philos. Trans. R. Soc. London **95** 65-87.

[9] Van Oss C J 1994 Interfacial forces in aqueous media. Marcel Dekker, New York.

[10] Shaw D J 1989 Introduction to colloid and surface chemistry. Butterworths, London.

[11] Bos R, Van der Mei H C and Busscher H J 1999 Physico-chemistry of initial microbial adhesive interactions – its mechanisms and methods for study. *FEMS Microbiol. Rev.* **23** 179-230.

[12] Jucker B A, Harms H and Zehnder J B. 1996 Adhesion of the positively charged bacterium *Stenotrophomonas (Xanthomonas) maltophilia* 70401 to glass and Teflon. *J. Bacteriol.* **178** 5472-5479.

[13] De Morais L C, Bernardes-Filho R and Assais O B G 2009 Wettability and bacteria attachment evaluation of multilayer proteases biofilm for sensor application. *World J. Microb. Biot.* **25** 123–129.

[14] Mazumder S, Falkinham J O, Dietrich A M and Puri I K 2010 Role of hydrophobicity in bacterial adherence to carbon nanostructures and biofilm formation. *Biofouling* **26** 333-339.

AUTHOR INDEX

Institute of Physics
Dirac House, Temple Back
Bristol BS1 6BE UK

CURRAN ASSOCIATES INC
proceedings.com